Triple C Model of Project Management

Communication, Cooperation, and Coordination

Industrial Innovation Series

Series Editor

Adedeji B. Badiru

Department of Systems and Engineering Management
Air Force Institute of Technology (AFIT) – Dayton, Ohio

PUBLISHED TITLES

Computational Economic Analysis for Engineering and Industry
Adedeji B. Badiru & Olufemi A. Omitaomu

Handbook of Industrial and Systems Engineering
Adedeji B. Badiru

Industrial Project Management: Concepts, Tools, and Techniques
Adedeji B. Badiru, Abidemi Badiru, and Adetokunboh Badiru

Systems Thinking: Coping with 21st Century Problems
John Turner Boardman & Brian J. Sauser

Techonomics: The Theory of Industrial Evolution
H. Lee Martin

Triple C Model of Project Management: Communication, Cooperation, Coordination
Adedeji B. Badiru

FORTHCOMING TITLES

Beyond Lean: Elements of a Successful Implementation
Rupy (Rapinder) Sawhney

Handbook of Industrial Engineerng Calculations and Practice
Adedeji B.Badiru & Olufemi A. Omitaomu

Handbook of Military Industrial Engineering
Adedeji B.Badiru & Marlin Thomas

Industrial Control Systems: Mathematical and Statistical Models and Techniques
Adedeji B. Badiru & Olufemi A. Omitaomu

Knowledge Discovery for Sensor Data
Auroop R. Ganguly, João Gama, Olufemi A. Omitaomu, Mohamed Medhat Gaber, and Ranga Raju Vatsavai

Modern Construction: Productive and Lean Practices
Lincoln Harding Forbes

Project Management: Systems, Principles, and Applications
Adedeji B. Badiru

Process Optimization for Industrial Quality Improvement
Ekepre Charles-Owaba and Adedeji B. Badiru

Research Project Management
Adedeji B. Badiru

Statistical Techniques for Project Control
Adedeji B. Badiru

STEP Project Management: Guide for Science, Technology, and Engineering Projects
Adedeji B. Badiru

Technology Transfer and Commercialization of Environmental Remediation Technology
Mark N. Goltz

Triple C Model of Project Management

Communication, Cooperation, and Coordination

Adedeji B. Badiru

CRC Press
Taylor & Francis Group
Boca Raton   London   New York

CRC Press is an imprint of the
Taylor & Francis Group, an **informa** business

CRC Press
Taylor & Francis Group
6000 Broken Sound Parkway NW, Suite 300
Boca Raton, FL 33487-2742

Library of Congress Cataloging-in-Publication Data

Badiru, Adedeji Bodunde, 1952-
    Triple C Model of Project Management: Communication, Cooperation, and Coordination / Adedeji B. Badiru.
        p. cm. -- (Industrial innovation series)
    Includes bibliographical references and index.
    ISBN 978-1-4200-5113-1 (alk. paper)
    1. Project management. I. Title. II. Series.

T56.8.B34 2008
658.4'04--dc22

2007044612

Visit the Taylor & Francis Web site at
http://www.taylorandfrancis.com

and the CRC Press Web site at
http://www.crcpress.com

Dedication

To Whom It May Concern.

Contents

Preface

TRIPLE C MODEL OF PROJECT MANAGEMENT COMMUNICATION, COOPERATION, AND COORDINATION

Everyone needs project management! The application of project management is vital in business, industry, government, and personal activities. Everyone and every organization needs project management because projects offer an avenue for the accomplishment of human effort. The core competencies that employers require of new employees often include leadership, team skills, and project management. So often is project management required in an organization that most now use management-by-projects (MBP) as a primary business strategy. Project management is the fastest growing profession because every effort and every other profession needs project management.

This book a part of the project management section of the Industrial Innovation book series by Taylor & Francis. The book presents the soft side of project management, which is not necessarily the easiest in managing large projects. People issues are fuzzy, ambiguous, and subject to emotional nuances and sentimental knee-jerk reactions. Consequently, the people side of project management requires more managerial care because no mathematical prescriptions are available to manage people. The book contains a comprehensive guide for the implementation of the Triple C Model of project management; with pragmatic integration of basic managerial concepts, tools, and techniques. The book has a good appeal in all types of human endeavor. A strategic plan cannot be realized until the process of project management is applied to implement the tactical objectives of the plan. The term "project management" generally implies the broad conceptual approaches used to manage projects within the constraints of time, cost, and performance expectations. Project management has proven very useful in different types of endeavors. Diverse areas such as engineering, construction, social work, health services, research, business, marketing, and education have benefited from the application of project management techniques. This book aims to (1) support increased worldwide use of project management, (2) show the applicability of project management to all human endeavors, and (3) present a guide using the Triple C model to implement project communication, cooperation, and coordination.

The book also presents the expanded body of knowledge for project management beyond what is typically covered in the primary body of knowledge introduced by the Project Management Institute. Chapters in the book cover management by projects, introduction to Triple C, communication, cooperation, and coordination. One chapter is devoted to Triple C case studies and another chapter contains presentation slides of case examples of the application of Triple C approach to real projects from industry. The last chapter in the book presents a template for project management implementation.

The Triple C model of project management will be beneficial for a variety of professional groups, including industrial engineers, management engineers, manufacturing engineers, operations research engineers, quality engineers, production managers, safety engineers, process improvement professionals, systems engineers, engineering consultants, engineering professors, and engineering directors. The primary lesson of the Triple C model presented in this book is not to take cooperation for granted. It must be pursued, solicited, and secured explicitly. The process of securing cooperation requires structured communication up front. It is only after cooperation is in effect that all project efforts can be coordinated.

Adedeji Badiru

Acknowledgments

My greatest thanks go to my wife, Iswat, for not only supporting this new project, but also helping to develop several of the graphic images contained in the book. Her spousal critique was an invluable resource. I would also like to thank my colleagues and associates at the Air Force Institute of Technology (Dayton, Ohio) for providing stimulating perspectives that led to new ways of looking at old ideas of managing projects. Finally, I would once again like to thank Cindy Carelli and her editorial and production staff for their continual commitment to excellence in coordinating book manuscripts in the Industrial Innovation book series of Taylor and Francis/ CRC Press.

Adedeji Badiru

Author Bio

Professor Badiru is professor and head of the department of Systems and Engineering Management at the Air Force Institute of Technology (AFIT). He was previously head of Industrial and Information Engineering at the University of Tennessee in Knoxville, and former professor of industrial engineering and dean of University College at the University of Oklahoma. He is a registered professional engineer and is a fellow of both the Institute of Industrial Engineers and the Nigerian Academy of Engineering. He holds a BS in industrial engineering, an MS in mathematics, an MS in industrial engineering from Tennessee Technological University, and a Ph.D. in industrial engineering from the University of Central Florida. His areas of expertise cover mathematical modeling, project modeling, analysis, management and control, economic analysis, productivity analysis, and improvement. He is the author of several books and technical journal articles. He is the editor of the *Handbook of Industrial and Systems Engineering* and is a member of several professional associations including the Institute of Industrial Engineers (IIE), the Society of Manufacturing Engineers (SME), the Institute for Operations Research and Management Science (INFORMS), the American Society for Engineering Education (ASEE), the New York Academy of Science, and the Project Management Institute (PMI).

1 Management by Project

We trained hard, but it seemed that every time we were beginning to form into teams, we would be reorganized. I was to learn later in life that we tend to meet any new situations by reorganizing; and what a wonderful method it can be for creating the illusion of progress while producing confusion, inefficiency, and demoralization.

**Actual source unknown, but often satirically attributed
to Petronius Arbiter, 210 BC**

If you want results, use project management! Every organization wants more results in less time with fewer resources. It is through the structured approach of project management that an organization can overcome confusion, inefficiency, and demoralization. Every organization needs project management! Everyone needs project management! The application of project management is vital in business, industry, government, and personal activities. Everyone and every organization needs project management because projects offer an avenue for the accomplishment of human effort. The core competence that employers require of new hires most often include *leadership, team skills,* and *project management.*

So often is project management required in an organization that most now use management by project (MBP) as a primary business strategy. The contemporary economy is built on service enterprises, such as information technology (IT) services, product design services, and supply chain. Many such services are conducted on a project basis. The formal definition below explains the importance of MBP in organizations.

Management by project (MBP) facilitates an application of the multidimensionality of factors that influence the accomplishment of organizational goals.

MBP has several benefits. It helps the process of learning leadership practices, team building, employee relations, interpersonal skills, and communication skills. Most projects have the same things in common:

- People issues
- Resource shortage
- Time crunch

Project principles, similarities, and practices are transferable across industries, across cultures, and across geographic boundaries. This makes MBP very versatile and generally applicable to different organization sizes, shapes, and locations. The vision and mission of a project will dictate the process of applying project management concepts, tools, and techniques in pursuit of organizational goals. The samples that follow show the need to create an operational platform for each project.

PROJECT VISION SAMPLE

In achieving the objectives of MBP, an organization must have a succinct vision for the project of interest and a clear mission statement to accomplish the project goals. A vision statement says how things ought to be for the future, while a mission statement indicates how things are currently. An example of a project vision statement is:

> "The shared vision of this project is to leverage all internal resources of ABICS Corporation to become the organization of choice to deliver ISO 9000 compliant engineering products and services globally whenever and wherever needed for now and the future."

PROJECT MISSION SAMPLE

In accordance with the vision statement, the project organization must develop a clear mission statement that adequately proclaims how the project will operate in order to achieve the accomplishment stated in the vision statement. An example of a project mission statement is:

> "The mission of the Triple C project is to provide environmentally conscious engineering services using integrated communication, cooperation, and coordination networking with our customers and suppliers; in accordance with contractual time, cost, and performance expectations. In particular, our mission encompasses:
>
> - Using project management techniques at the outset of each project
> - Opening and maintaining a communication line between our engineers and customer's representatives
> - Forming an alliance with leading edge subcontractors
> - Enhancing our delivery potential with the infusion of information technology tools"

QUALITATIVE SIDE OF PROJECT MANAGEMENT

This book presents the soft side of project management, which is not necessarily the easiest. People issues are ambiguous and subject to emotional nuances and sentimental knee-jerk reactions. Consequently, the people side of project management requires more managerial care because no mathematical prescriptions are available to manage people. This book contains a comprehensive guide for the implementation of the Triple C model of project management, with pragmatic integration of basic managerial concepts, tools, and techniques. The contents of this book are applicable to all types of human endeavor. A strategic plan cannot be realized until the process of project management is applied to implement the tactical objectives of the plan. The term "project management" generally implies the broad conceptual approach used to manage projects within the constraints of time, cost, and performance expectations. Project management has proved very useful in different types of endeavors. Diverse areas such as engineering, construction, social work, health

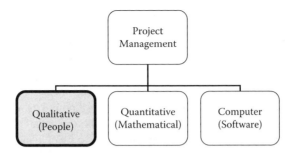

FIGURE 1.1   Role of qualitative focus in project management.

services, research, business, marketing, and education have benefited from the application of project management techniques.

The primary implementations of project management center around qualitative, quantitative, and computer techniques. Each of these has been extensively addressed independently and collectively in many publications. A common mistake is to quickly jump to computer implementation because computer tools for project management have become very accessible in recent years. The fact is that computer implementations cannot succeed without appropriate qualitative analysis and quantitative modeling. The qualitative and people aspects of project management, constituting the focus of this book, are highlighted in Figure 1.1. It is a strong belief of the author that people still run projects regardless of the level of quantitative modeling and computer tools available. Project success rests on the "shoulders" of people—people as project team members, people as project owners, or people as project customers. Qualitative analysis must precede quantitative analysis before computer analysis can succeed.

The techniques of project management have become major tools used to accomplish goals and objectives. Project management, as a body of knowledge, is reshaping business processes. Management-by-projects offers a huge competitive advantage for companies. But the potential is not yet fully utilized across organizations.

CORPORATE STRATEGY

Project management is evolving more and more as a corporate entity. The characterization of a project as a corporate entity embodies several elements of the usual project management body of knowledge as defined by the Project Management Institute. In the context of management, "corporate" means "organization." Thus, corporate entrepreneurship is akin to "organizational intrapreneurship," which is the process of utilizing or leveraging internal organizational resources, skills, and expertise to generate innovative products, services, and enhanced processes for the organization. Intrapreneurship was the process used by 3M to develop the popular Post-It note pads. So, an organization may want to market its products based on the syntactic preferences of the organization. All these words refer to the same thing—organizational process improvement, regardless of whether it is in military or civilian organizations.

OGA IN WORLD BUSINESS LEADERSHIP

Project management is bearing fruits for businesses. Structured project management gives an organization the ability to stay One Generation Ahead (OGA) of the competition. The term OGA represents a play on the Nigerian word "oga" meaning "leader" or "boss." This is aptly appropriate for how project management, if applied enterprise-wide, can help an organization achieve leadership position in business and industry. The section title, "OGA in Business" implies that the organization has achieved a level of business leadership through the application of project management.

The practice of project management continues to grow at an astronomical rate. This growth is evidenced by the present membership size of the Project Management Institute (PMI), the international professional organization for project management. The global appeal and relevance of project management has created many professional opportunities for practitioners, consultants, and researchers. This leads to the need for simple and practical guide books. This book presents the Triple C model of project management, which encompasses the communication, cooperation, and coordination requirements in any project.

As an example, the Forbes Global CEO Conference (September 9–12, 2007, Singapore) emphasized the role of project management in securing global business leadership for an organization. The conference, hosted by Forbes magazine, had over 400 chief executive officers (CEOs) and business leaders in attendance from all around the world.

The theme for the conference was "Drivers of Success," and project management was identified as the key factor that helps organizations to achieve predictable, repeatable, and profitable results.

At the Forbes conference, the executive director of the Project Management Institute (PMI), Mr. Greg Balestrero, emphasized that to stay competitive, businesses must innovate and adapt to change. "Innovation is not optional," he said quoting *Dealing with Darwin* author Geoffrey Moore. "To survive, organizations must gain stakeholder confidence and support by demonstrating an ability to innovate and deal with cyclical inertia."

He went on to say that through the discipline of project management, companies can meet the needs of customers by delivering on promises predictably and repeatedly. Mr. Balestrero also commented on the growing need for companies to deliver quality and safe products and how project management standards can be applied to improving *planning, coordination, communication,* and production quality controls. These requirements are precisely the bastions of the Triple C model as presented in this book. As more and more profitable project management results are announced by worldwide business leaders, more and more organizations will embrace MBP as a core business strategy.

TIME-COST-PERFORMANCE TRADE-OFFS

Triple C is different from (but related to) the usual concept of triple constraints of time, cost, and performance. Figure 1.2 presents the basic triple constraints axial relationship between time, cost, and performance, while Figure 1.3 illustrates the

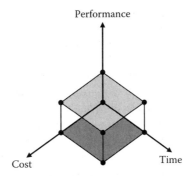

FIGURE 1.2   Basic triple constraints relationship.

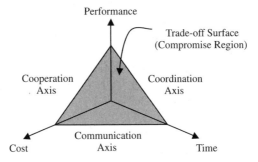

FIGURE 1.3   Triple C model mapped over triple constraints.

differences and the relationships when Triple C is mapped over the triple constraints. Effective accomplishments of time, cost, and performance can only be sustained with effective communication, cooperation, and coordination. In Figure 1.2, the time, cost, and performance relationship is presented in a framed box. We are often advised to "think outside the box" in order to develop creative and innovative solutions. However, in a time-cost-performance-constrained environment, we should "operate within the box" in order to comply with project requirements. It is from within the box that we can stick our necks out to identify creative and innovative solutions to project challenges, particularly those requiring trade-offs and compromise. Inference: Think outside the box, but operate from within it!

DEFINITION OF PROJECT MANAGEMENT

The traditional (age-old) definition of a project says that

"A project is a unique endeavor that has a definite beginning and a definite end."

This is a limiting definition since it does not take into account the myriad of factors that influence contemporary projects. Modern projects are no longer limited to planning and control of a collection of activities. While control is a fundamental managerial activity in executing a project, the traditional form of control is no longer effective.

Control can be defined as the set of actions (proactive, reactive, or corrective) that are designed to motivate team members to perform in a way that is consistent with organizational goals and objectives. For contemporary projects, traditional formal control processes should be redesigned to be more adaptive to the changing environment within which a project is executed. The conventional predetermined rules, policies, and procedures should be modified to accommodate more flexible team empowering processes. Strict hierarchical supervisor-subordinate control relationships do not work well with modern knowledge workers, who are more sophisticated and discerning in their job choices than their predecessors, who had allegiance to task and activity management structures.

Modern projects encompass a lot more than activity management. This author expanded the conventional view of a project by offering an alternate definition of project management, which says

"Project management is the process of managing, allocating, and timing resources to achieve a given goal in an efficient and expeditious manner."

This expanded definition permits the incorporation of swiftness and expediency to the conventional perception of project management. Project goals are achieved through an integrative synergy between people, tools, and process. As the integral component of project efforts, human resources should enjoy a high priority in an organization's strategic plans. If an organization treats its workforce with dignity, it will bring out the best in the workers, both individually and as a group. Project life cycle revolves around several factors that define the overall utility of the project. Figure 1.4 shows selected examples of the potential additional dimensions of contemporary projects in addition to the traditional cost, time, and performance considerations.

Unlike traditional projects, contemporary projects are evaluated on the basis of several factors ranging from being useful, beneficial, practical, appropriate, customer oriented, cost effective, performance driven, time sensitive, necessary, to being human centric.

MANUFACTURING DEFINED AS A PROJECT

The author's first book on project management was a 1988 publication on project management in manufacturing. The initial proposal for the book was a hard sell to the publisher because all the initial review reports were negative. The so-called "gurus" of project management at that time, who reviewed the proposal, commented that "manufacturing" was not amenable to the application of project management because it did not fit the conventional definition of a project as "a unique one-of-a-kind endeavor with a definite beginning and a definite end." After reflecting on the negative comments for several days, the author came back with a counterpoint that stressed that manufacturing could, indeed, be viewed as a series of contiguous projects, whereby each production cycle is a "unique one-of-a-kind endeavor." That counterpoint was what sold the proposal to the publisher. The book was published successfully. It was the first book of its kind at that time. In fact, the publisher characterized the book as being ahead of its time. The reviewers who initially thought it

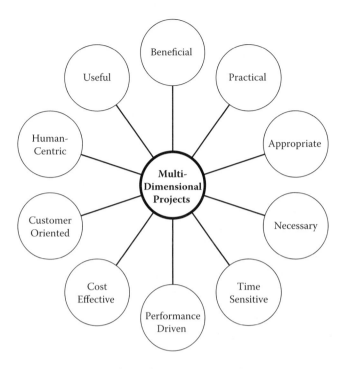

FIGURE 1.4 Multidimensional views of contemporary projects.

was a bad idea ended up fully supporting the project. As we all know now, project management is widely applied to manufacturing activities. It can be equally applied to any other organizational endeavor.

PEOPLE COMPONENT OF PROJECT MANAGEMENT

Because people run projects, the individual attributes of project team members are important for success. Organizations thrive by investing in three primary resources: the people who do the work, the tools that the people use to do the work, and the process that governs the work that the people do. Of the three, investing in people is the easiest thing an organization can do, and we should do it whenever we have an opportunity. First and foremost is the need to assess the personal ethics of a team member. Figure 1.5 shows the components of personal ethics. Work ethics relate to how a person dedicates himself or herself to the task at hand. It is quite possible to have a highly qualified and competent individual with very low work ethics. In that case, a project many not be able to get much work out of the individual. The communication and cooperation processes of Triple C can help mitigate the adverse effects of low work ethics. This is done through statement of clear objectives, explanation of the importance of each person's role, assignment to team expectations, and empowerment to get things done. In most cases, given the challenge and opportunity to thrive in an organization, those with low work ethics can still be rehabilitated for the benefit of the project.

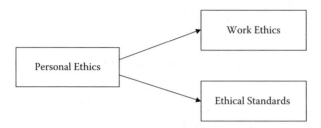

FIGURE 1.5 Distribution of personal ethics.

Ethical standards relate to an individual's sense of integrity and credibility. This infers a person's conscious and deliberate commitment to play by the rules. Even when rules have to be bent to get things done, it should be done within the sphere of honesty, honor, and justifiable reasoning.

Figure 1.6 shows some factors that influence worker productivity on a project. The project leadership must recognize signals of these factors when attempting to coordinate project efforts. The overall wellness of an employee directly affects productivity. Factors such as work stress, home stress, social stress, safety practices, personal finances, social consciousness, and family situations do affect wellness and work productivity. To not recognize these issues in a project environment is to pretend that human frailty and flaws don't exist. A healthy and happy worker is a more productive worker. Health issues are not always physical infirmities that can be perceived on the exterior. In fact, most afflictions in the work force are subtle. Communication and cooperation, following the Triple C approach, are essential for eliciting and dealing with them.

Figure 1.7 further illustrates the importance of overall wellness of a worker in terms of physical, social, emotional, and spiritual needs of people. When these needs

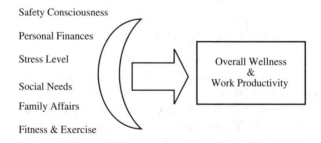

FIGURE 1.6 Factors influencing worker productivity on a project.

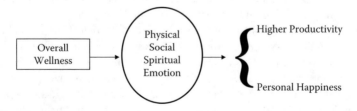

FIGURE 1.7 Overall Wellness as a Function of Personal Needs.

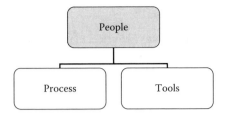

FIGURE 1.8 Harmony of people, process, and tools.

are met, workers tend to be happier, more productive, and more *cooperative*. This makes it possible for principles of Triple C to be effective in the project environment.

The three components of a project are illustrated in Figure 1.8, where people, process, and tools must work in consonance and have harmonious integration to achieve project objectives. To improve project performance, the leadership must invest in advancing the people (through training), improving the process (through work flow assessment), and update tools (through procurement strategies).

PROJECT LEADERSHIP

The quality of leadership available to a project is of utmost importance in moving the project toward success. Leadership is an intrinsic quality that represents a combination of several personal attributes. Some of the attributes can the learned, while others are innate. Naturally, the project manager is expected to provide leadership for the project. A leader should always be positive and enthusiastic about the mission of the project at hand. A leader should always present a message positively to rally the project "troops" toward the intended goals of the project. A leader should always find the silver lining in a gloomy scenario. To miss the silver lining is to pave the way for losing the support of team members. Other essential criteria for an effective project manager are:

- Technical proficiency
- Critical thinking skills
- Communication skills
- Cooperation skills
- Coordination skills
- Resource management skills
- Leadership (mission-focused and vision-centric)

Notice that three of the seven elements of the criteria are aligned with the requirements of the Triple C approach. In summary, effective leadership has the following attributes:

1. Leadership must show results.
2. Leadership must lead by results.
3. Results facilitate cooperation.

JACK WELCH ON LEADERSHIP

Jack Welch, Former CEO of General Electric, regarded by many as the greatest CEO of our lifetime, once discussed his leadership philosophy of *"Four E's Wrapped in a P."* He stated that this philosophy enabled all of GE's employees to change. So what are the four E's, and the P?

> **E: Energy!** Being energetic allows you to accomplish what needs to be done, when it needs to be done!
> **E: Energize others!** If you don't energize others, your energy is wasted!
> **E: Edge!** Have what it takes to be decisive. Stand out, exude confidence in your power and your position!
> **E: Execute!** Sit down, plan it out, stand up, and DELIVER!
> **And the P?**
> **PASSION!!** Have passion in what you do, and what you believe, because it is with this passion that you will draw the power for all the E's!

AXIOM OF PROJECT LEADERSHIP

The axiom of leadership can be seen in the verse below, which describes the essential and preferred attributes of a leader.

BE A PROJECT LEADER, NOT A BOSS

The boss drives his project
The leader inspires his project team

The boss depends on project authority
The leader depends on goodwill

The boss evokes fear
The leader radiates affection

The boss says "I"
The leader says "We"

The boss points to who is wrong
The leader identifies what is wrong

The boss knows how it is done
The leader shows how to do it

The boss demands respect
The leader earns respect

The boss dictates his desires
The leader involves his team

So, be a project leader
Instead of just being a boss.

SECURING MANAGEMENT SUPPORT

Every project needs management support. It is a fact that much is made of the importance of management support. But the reality is that many of us don't know how to secure management support. Confronting and "pressing" management is not the way to get management support. A project must be justified and "sold" to the management. Management must be convinced with facts regarding the merits of the project.

There is nothing to be gained by standing up to management. You can suck up to management without rolling over. In fact, more can be gained by working with management than using a confrontational approach. Using the Triple C approach, the project team can secure management support much more easily than using the conventional approach of ridiculing management. To resolve not to work with management due to management ineptitude would be counterproductive and self-effacing. In the final analysis, each person will be judged by the substance of his or her work; not by what support is not provided by management. Successful project managers face the management obstacles, but they still find ways to thrive in their project responsibilities. The same conciliatory approach that Triple C advocates for working with management and team members can be applied to other non-work-related interactions.

BASIC ELEMENTS OF PROJECT MANAGEMENT

Although there are many intricate steps in project management, the basic elements are summarized below and will be addressed in various sections of the book:

1. Planning
2. Organizing
3. Scheduling
4. Tracking and control

The expanded steps of project management are shown in Figure 1.9. All the elements in the figure will be addressed in subsequent sections as they relate to the implementation of Triple C model. The key aspect of using Triple C across the steps is to realize that a contemporary project must be adaptive and flexible in permitting iterative interactions of the steps. The steps are not rigid or cast in stone. This means that problem definition and redefinition must be possible as the project unfolds. The mission statement (or charter) must be revisited and modified as project realities evolve. Iterative planning must be permitted as project structure develops. Resources must be permitted to flow (in and out) at various stages of the project as project needs develop. Tracking, reporting, and control should be embedded all along the hierarchy of project steps. "Coup de grace" or termination (pulling the plug) should be an option at every stage of the project.

Projects face the greatest risk of failure right at the beginning and at the end. Failure to properly define and initiate a project sets the tone for later failure. Many projects start in a failure mode because they are ill-conceived and poorly initiated. The other critical point of failure for a project is at the end when there is a lack of

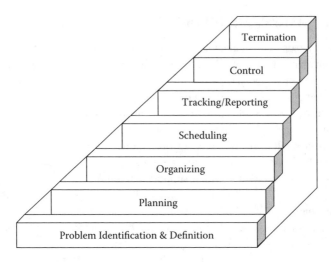

FIGURE 1.9   Expanded steps of project management.

resolute phase-out. A project must be phased out when it needs to be phased out. The sunset clause principle below explains this further.

SUNSET CLAUSE IN PROJECT AGREEMENTS

A sunset clause is derived from a legal basis that automatically repeals or terminates all or portions of a statute or regulation after a specific date. Further legislative action is needed to extend the statute or regulation. A law that does not have a sunset clause goes on indefinitely. For project management purposes, symbolic sunset clauses are important to ensure that a project agreement terminates after the period initially covered by the agreement. The origin of the sunset clause can be traced to the Roman principle of "*Ad tempus concessa post tempus censetur denegata*," which translates to "what is admitted for a period will be refused after the period." Thus, a project agreement that is developed for a specific period is valid only for that period. Without a sunset clause, projects that deserve to "die" may live on indefinitely, consuming scarce resources and disrupting productive flow of other projects. Collaboration and cooperation built up on projects that are phased out should be transferred to other projects, rather than being used to fuel indefinite existence of projects that have outlived their usefulness.

PROJECT PLANNING

"A plan is the map of the wise."

Adedeji Badiru, 1993

Planning is the roadmap that links goals to actions through the components of the work breakdown structure of a project. Figure 1.10 illustrates how the linkage is accomplished. It is essential to identify and resolve conflicts in project planning

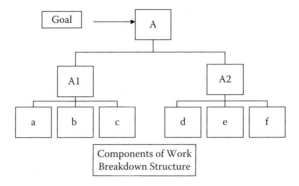

FIGURE 1.10   Project work breakdown structure.

early before resources are committed to work elements that do not add value to the pursuit of an organization's goals.

THE ABILENE PARADOX: CONFLICT IN GROUP AGREEMENT

A classic example of conflict in project planning is illustrated by the *Abilene Paradox* (Harvey, 1988). The text of the paradox, as presented by Harvey, is summarized below.

It was a July afternoon in Coleman, a tiny Texas town. It was a hot afternoon. The wind was blowing fine-grained West Texas topsoil through the house. Despite the harsh weather, the afternoon was still tolerable and potentially enjoyable. There was a fan blowing on the back porch; there was cold lemonade; and finally, there was entertainment: dominoes. Perfect for the conditions. The game required little more physical exertion than an occasional mumbled comment, "shuffle 'em," and an unhurried movement of the arm to place the spots in the appropriate position on the table. All in all, it had the makeup of an agreeable Sunday afternoon in Coleman until Jerry's father-in-law suddenly said, "Let's get in the car and go to Abilene and have dinner at the cafeteria."

Jerry thought, "What, go to Abilene? Fifty-three miles? In this dust storm and heat? And in an un-air-conditioned 1958 Buick?" But Jerry's wife chimed in with, "Sounds like a great idea. I'd like to go. How about you, Jerry?" Since Jerry's own preferences were obviously out of step with the rest, he replied, "Sounds good to me," and added, "I just hope your mother wants to go."

"Of course I want to go," said Jerry's mother-in-law. "I haven't been to Abilene in a long time." So into the car and off to Abilene they went. Jerry's predictions were fulfilled. The heat was brutal. The group was coated with a fine layer of dust that was cemented with perspiration by the time they arrived. The food at the cafeteria provided first-rate testimonial material for antacid commercials.

Some four hours and 106 miles later, they returned to Coleman, hot and exhausted. They sat in front of the fan for a long time in silence. Then, both to be sociable and to break the silence, Jerry said, "It was a great trip, wasn't it?" No one spoke. Finally, his father-in-law said, with some irritation, "Well, to tell the truth, I really didn't enjoy it much and would rather have stayed here. I just went along because the three of you were so enthusiastic about going. I wouldn't have gone if you all hadn't pressured me into it."

Jerry couldn't believe what he just heard. "What do you mean 'you all'?" he said. "Don't put me in the 'you all' group. I was delighted to be doing what we were doing. I didn't want to go. I only went to satisfy the rest of you. You're the culprits." Jerry's wife looked shocked. "Don't call me a culprit. You and Daddy and Mama were the ones who wanted to go. I just went along to be sociable and to keep you happy. I would have had to be crazy to want to go out in heat like that."

Her father entered the conversation abruptly. "Hell!" he said. He proceeded to expand on what was already absolutely clear. "Listen, I never wanted to go to Abilene. I just thought you might be bored. You visit so seldom I wanted to be sure you enjoyed it. I would have preferred to play another game of dominoes and eat leftovers in the icebox."

After the outburst of recrimination, they all sat back in silence. There they were, four reasonable sensible people who, of their own volition, had just taken a 106-mile trip across a godforsaken desert in a furnace-like temperature through a cloud-like dust storm to eat unpalatable food at a hole-in-the-wall cafeteria in Abilene, when none of them had really wanted to go. In fact, to be more accurate, they'd done just the opposite of what they wanted to do. The whole situation simply didn't make sense. It was a paradox of agreement."

This example illustrates a problem that can be found in many organizations or project environments. Organizations often take actions that totally contradict their stated goals and objectives. They do the opposite of what they really want to do. For most organizations, the adverse effects of such diversion, measured in terms of human distress and economic loss, can be immense. A family group that experiences the Abilene paradox would soon get over the distress, but for an organization engaged in a competitive market, the distress may last a very long time. Six specific organizational ills are revealed by the paradox:

1. Organization members agree privately, as individuals, as to the nature of the situation or problem facing the organization.
2. Organization members agree privately, as individuals, as to the steps that would be required to cope with the situation or solve the problem they face.
3. Organization members fail to accurately communicate their desires and/or beliefs to one another. In fact, they do just the opposite and, thereby, lead one another into misinterpreting the intentions of others. They misperceive the collective reality. Members often communicate inaccurate data (e.g., "Yes, I agree"; "I see no problem with that"; "I support it") to other members of the organization. No one wants to be the lone dissenting voice in the group.
4. With such invalid and inaccurate information, organization members make collective decisions that lead them to take actions contrary to what they want to do and, thereby, arrive at results that are counterproductive to the

organization's intent and purposes. For example, the Abilene group went to Abilene when it preferred to do something else.

5. As a result of taking actions that are counterproductive, organization members experience frustration, anger, irritation, and dissatisfaction with their organization. They form subgroups with supposedly trusted individuals and blame other subgroups for the organization's problems.

6. The cycle of the Abilene paradox repeats itself with increasing intensity if the organization members do not learn to manage their agreement.

We have witnessed many project situations where, in private conversations, individuals express their discontent about a project and yet fail to repeat their statements in a group setting. Consequently, other members are never aware of the dissenting opinions. In large organizations, the Triple C model can help in managing communication, cooperation, and coordination functions to avoid the Abilene paradox. The lessons to be learned from proper approaches to project planning can help avoid unwilling trips to Abilene. Specifically, the use of Triple C communication and cooperation approaches can help mitigate paradoxical inconsistencies.

PROJECT ORGANIZING

Organizational resources are mapped against project requirements in order to achieve goals. A requirement of this phase of project management is to have an effective organizational structure. An outline of respective responsibilities of personnel and other resources is required to avoid confusion, misinterpretation, and ambiguity. A common tool for organizing a project is the responsibility matrix, an example of which is shown in Figure 1.11. The matrix addresses who does what, when, and with which resource across different projects. It also addresses resource estimates and resource availability. Defining the lines of responsibility explicitly helps to avoid, minimize, or mitigate ambiguities in project execution.

PROJECT SCHEDULING

Scheduling involves establishing the precedence structure and execution relationships among activities in a project. The basic structure is represented in terms of a network of activities as shown in Figure 1.12. Scheduling is often recognized as the major function in project management. The main purpose of scheduling is to allocate resources so that the overall project objectives are achieved within a reasonable time span. In general, scheduling involves the assignment of time periods to specific tasks within the work schedule. Resource availability, time limitations, urgency level, required performance level, precedence requirements, work priorities, technical constraints, and other factors complicate the scheduling process. Thus, the assignment of a time slot to a task does not necessarily ensure that the task will be performed satisfactorily in accordance with the schedule. Consequently, careful control must be developed and maintained throughout the project scheduling process.

Who does what when with which resource?

    Resource Estimate: How Much?

    Resource Availability: When?

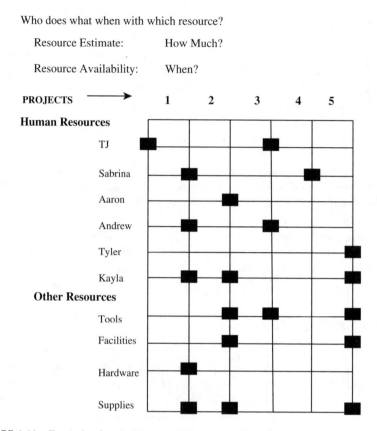

FIGURE 1.11   Example of project responsibility matrix.

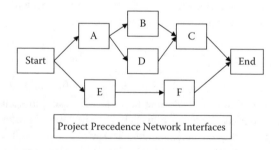

FIGURE 1.12   Network representation of a project.

PROJECT TRACKING, REPORTING, AND CONTROL

Tracking, monitoring, and control of activities and events are typically accomplished by using a Gantt chart, which uses timelines to depict the when and how of each activity. Milestones are also indicated in the tracking chart. The tracking and reporting phase involves the process of checking whether or not project results conform to

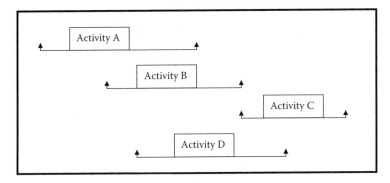

FIGURE 1.13   Project monitoring and event schedule.

plans and specifications. Tracking and reporting are prerequisites for project control. A properly organized report of project status will quickly identify the deficiencies in the progress of the project and help pinpoint necessary corrective actions. In the project control function, necessary actions are taken to correct unacceptable deviations from expected performance. Control is effected by measurement, evaluation, and corrective action. Measurement is the process of measuring the relationship between planned performance and actual performance with respect to project objectives. The variable to be measured, the measurement scale, and measuring approach should be clearly specified during the planning stage. Figure 1.13 presents a basic structure for tracking activities. The same tools typically used in Six Sigma define, measure, analyze, improve, and control (DMAIC) steps are applicable to project monitoring and control.

PROJECT EXIT PLAN

Every project should have an exit plan. An organization should know when to let a project go. If the signs of fruitlessness are there, a bad project should not be allowed to continue. A project that has exceeded its useful life should be phased out. Project phase-out or termination is the last stage of a project. The phase-out of a project is as important as the initiation of the project. There should be as much commitment to the termination of a project at the appropriate time as the commitment to initiating the project. A project should not be allowed to drag on needlessly after the expected completion time. The popularity of a project should not constitute a blank check of continuity. Popularity does not ensure that a project will continue to deliver. So, a terminal activity should be defined for a project during the project planning phase. An example of a terminal activity may be the submission of a final report, the "power-on" of new equipment, or the signing of a release order. The conclusion of such an activity should be viewed as the completion of the project, even though that completion may signify the beginning of the next project phase, cycle, or new project.

To prevent a project from dragging on needlessly, definite arrangements should be made about when the project should end. However, provisions should be made for follow-up activities or projects that may further improve the results of the initial project. These follow-up or spin-off projects should be managed as totally separate projects but with proper input-output relationships within the sequence of projects. If a project is not terminated when appropriate, the motivation for it will wane and subsequent activities may become counterproductive. This is particularly true for technology-based projects where the "fear of the unknown" and "resistance to change" are already major obstacles.

CONSEQUENCES OF PROJECT FAILURE: THE CODE OF HAMMURABI

The importance of project management goes back several centuries. In ancient times, project failures had grave consequences; literally "an eye for an eye." The Code of Hammurabi (ca. 1760 BC) typifies grave consequences for project failures. In the modern era, consequences may not exceed penalties and loss of future opportunities. In the time of Hammurabi, loss of life could be a consequence of project failure. The code contained promulgations such as the following:

> If a builder build a house for some one, and does not construct it properly, and the house which he built fall in and kill its owner, then that builder shall be put to death.

> If it kill the son of the owner the son of that builder shall be put to death.

> If it kill a slave of the owner, then he shall pay slave for slave to the owner of the house.

> If it ruin goods, he shall make compensation for all that has been ruined, and inasmuch as he did not construct properly this house which he built and it fell, he shall re-erect the house from his own means.

> If a builder build a house for some one, even though he has not yet completed it; if then the walls seem toppling, the builder must make the walls solid from his own means.

As can be seen from the above examples, the sense of being accountable for project failures has been around for a long time. Even though the consequences are no longer as grave, the need to preempt failures is not any less. Even nowadays, accidental or deliberate project failures can lead to catastrophes and loss of lives. The use of proper planning, organizing, scheduling, and tracking/control are always very essential. The concepts presented in this book are used in accomplishing these requirements.

Figure 1.14 shows a summarization of project management requirement into three rules for communication, cooperation, and coordination. Following an expert systems type of approach, the rules can be viewed as summarized in "Rules of Triple C" in Figure 1.14.

Figure 1.14 affirms that communication is the root of success for any project. Figure 1.15 shows a diagram of the Triple C rules as basic stages of project management.

PROJECT MANAGEMENT KNOWLEDGE AREAS

The Project Management Institute (PMI) has identified nine major functional areas that embody the practice of project management. These are compiled into the *Project*

> **Rules of Triple C**
>
> *If* **communication** *is done properly,*
>
>   *then* **cooperation** *will occur intrinsically;*
>
> *If* **cooperation** *exists, then* **coordination** *will be effective;*
>
> *If* **coordination** *is in place, then project success is assured.*

FIGURE 1.14   Triple C expert system rules.

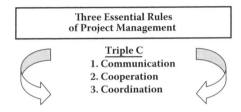

FIGURE 1.15   Essential rules of project management.

Management Book of Knowledge (PMBOK) by the Institute. The knowledge areas are scope, cost, schedule, risk, communications, human resources, procurement, quality, and integration. These are shown in Figure 1.16.

The last area covers project integration, ethics, health, safety, welfare issues, and professional responsibility. These are topics that are traditionally not addressed explicitly in project management. Several topics covered in this book address many of the key elements embodied in the knowledge areas. Scope management specifies the project charter, project plan, and relationship to other projects going on in an organization. Project integration and scoping are covered in this chapter. Subsequent chapters present topics on time management, resource management, risk management, quality and process management, communications management, and cost management. Although there are nine distinct areas presented in the PMBOK, all the requirements can be covered in terms of the basic three constraints on any project, as presented earlier in Chapter 1. The triple constraints of time, cost, and performance are shown again below in terms of the first three elements of PMBOK:

- Scope (performance)
     → Performance specs, output targets, etc.
- Schedule (time)
     → Due date expectations, milestones, etc.
- Cost (budget)
     → Budget limitations, cost estimates, etc.

If the above three elements are managed effectively, all the other areas will be implicitly or explicitly covered. Cost and schedule are subject to risk. Communications

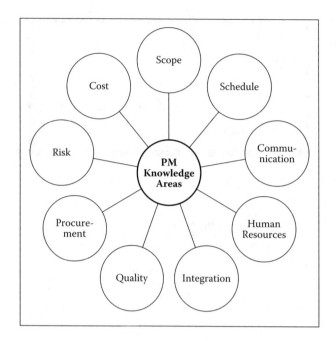

FIGURE 1.16 Project management knowledge areas.

are required for effective scoping. Human resources affect scope, cost, and schedule. Procurement provides the tools and infrastructure for project delivery. Quality implies performance and vice versa. Integration creates synergy, accountability, and connection among all the elements. Project planning is the basis for achieving adequate attention to all the requirements of a project. The larger and more complex a project, the more critical the need for using structured project planning.

PARADOX OF PROJECT KNOWLEDGE

PMBOK attempts to map PMI's own project management lexicon over business practices and inherent operations in the project environment. But in spite of the number of certified project management professionals around the world, many projects still fail. Project success requires knowledge, skills, tools, and techniques. Certification confirms knowledge, but does not necessarily translate to operational implementation. Operational implementation requires interpersonal skills and techniques to coordinate human efforts to get the job done. That is, the processes of human communication and cooperation are required to get the job done and ensure project success. The caveat of each knowledge element in PMBOK is that it must be implemented by humans. Humans need communication, cooperation, and coordination.

PROJECT PARTNERING

Project partnering is an effective way to share resources and transfer learning from one project to another. Technology transfer processes help to transfer tools and

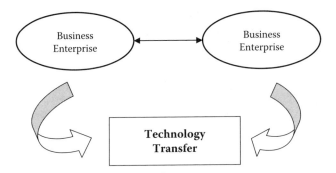

FIGURE 1.17   Project technology transfer model.

techniques from research or academic institutions to business enterprises. Figure 1.17 shows a possible model for knowledge transfer between project entities.

Technology transfer for project management purposes can be achieved in various forms. Three technology transfer modes are presented here to illustrate basic strategies for getting a project technology or knowledge from a source to a target point. Technology can be transferred in one or a combination of the following strategies:

1. Transfer of complete project products
2. Transfer of project procedures and guidelines
3. Transfer of project concepts, theories, and ideas

Transfer of Complete Products

In this case, a fully developed product is transferred from a source to a target. Very little product development effort is carried out at the receiving point. However, information about the operations of the product is fed back to the source so that necessary product enhancements can be pursued. So, the technology recipient generates product information, which facilitates further improvement at the technology source. This is the easiest mode of technology transfer and the most discernible.

Transfer of Project Procedures and Guidelines

In this technology transfer mode, procedures (e.g., blueprints) and guidelines are transferred from a source to a target. The technology blueprints are implemented locally to generate the desired services and products. The use of local raw materials and manpower is encouraged for the local production. Under this mode, the implementation of the transferred technology procedures can generate new operating procedures that can be fed back to enhance the original technology. With this symbiotic arrangement, a loop system is created whereby both the transferring and the receiving organizations derive useful benefits.

Transfer of Technology Concepts, Theories, and Ideas

This strategy involves the transfer of the basic concepts, theories, and ideas supporting a given technology. The transferred elements can then be enhanced, modified, or customized within local constraints to generate new technology products. The local modifications and enhancements have the potential to generate an identical technology, a new related technology, or a new set of technology concepts, theories, and ideas. These derived products may then be transferred back to the original technology source; a kind of reverse outsourcing. Transferred technology must be implemented to work within local limitations. Local innovation, patriotism, dedication, and cultural flexibility to adapt are required to make outsourcing technology transfer successful.

PROJECT ANECDOTES

Anecdotal examples, axioms, proverbs, adages, maxims, and aphorisms are essential for committing project lessons to memory. While some anecdotes are sarcastic, many do present accurate representations of project realities. Some common examples found in conventional project management circles are presented in this section. They offer a sort of comic relief for the common stress of managing complex projects.

AN ENGINEER'S DECLARATION OF A PROJECT

Project management involves various interfaces within an organization. The exchange of information at each interface is crucial to the success of a project. The declaration below, the original source of which is unknown, takes a humorous look at the importance of information transfer and feedback in a project environment. It has been circulated informally among project teams for many years. The original composition has been modified here, both in content and language, to make it suitable for publication:

> In the beginning there was the Project. With the Project, there was a Plan and a Specification. But the Plan was without form and the Specification was void. Thus, there was darkness upon the faces of the Engineers.
>
> The Engineers, therefore, spoke unto their Project Leaders, "this is a crock of crap and we cannot abide the stink that abounds."
>
> And the Project Leaders spoke unto their Unit Managers, "this is a crock of waste and we cannot abide the odor which abounds."
>
> And the Unit Managers spoke unto their Sub-Section Managers, "this is a vessel of waste and the odor is very offensive."
>
> And the Sub-Section Managers spoke unto their Section Managers, "this vessel is full of that which makes things grow and the characteristics thereof are exceedingly strong."
>
> And the Section Managers spoke unto the General Manager, "the contents of this vessel are very powerful and will promote strong growth of the Company."
>
> And the General Manager looked at the Project and saw that it was good. He, therefore, declared the Project fit for shareholders' consumption.

PROJECT MANAGEMENT PROVERBS

The stress of managing large projects often calls for proverbs that aid project personnel in seeing the lighter sides of their functions. The proverbs below represent a small sample of the various proverbs and sayings typically found in engineering and management circles.

The same work under the same conditions will be estimated differently by ten different estimators or by one at ten different times.

Construction engineer's mind is like concrete—thoroughly mixed up, and permanently set.

You can bamboozle an engineer into committing to an unreasonable deadline, but you can't con him into meeting the deadline.

The more ridiculous the deadline, the more it costs to try to meet it.

The more desperate the situation, the more optimistic the engineer.

Too few engineers on a project can't solve the problems; too many create more problems than they can solve.

You can freeze the users' specifications, but you can't stop them from expecting.

The conditions of a promise are forgotten whenever the promise is remembered.

What you don't know about your project is what really hurts you.

A user will tell you only what you ask about, and nothing more.

What is not on paper has not been said or heard.

No large project is ever installed on time, within budget, with the same staff that started it.

Projects progress quickly until they become 90 percent complete; then they remain there forever.

The rate of change of engineering projects often exceeds the rate of progress.

Debugging engineering systems creates new bugs that are unknown to engineers.

Progress reports are intended to show the lack of progress.

Murphy is alive and well in every project.

Peter's principle prevails in every organization.

Parkinson's law is every engineer's favorite.

PROJECT ENGINEER'S PHRASES

Engineers use trademark phrases to convey ideas when dealing with project managers and clients. The phrases below offer hilarious interpretations of how project engineers communicate. A project manager must be able to read in between the lines to get an accurate picture of the status of a project. When a project is declared as being complete, it may actually mean that the implementation stage is about to begin.

Phrase 1: "The concept was developed after years of intensive research."
Meaning: It was discovered by accident.

Phrase 2: "The design will be finalized in the next reporting period."
Meaning: We haven't started this job yet, but we've got to keep the manager happy.

Phrase 3: "A number of different approaches are being tried."
Meaning: We don't know where we're going yet, but we're moving.

Phrase 4: "The project is slightly behind schedule due to unforeseen difficulties."
Meaning: We are working on something else.

Phrase 5: "We have a close project coordination."
Meaning: Each project group does not know what the others are doing.

Phrase 6: "An extensive report is being prepared on a fresh approach to the problem."
Meaning: We just hired three guys ... It will take them a while to figure out the problem.

Phrase 7: "We've just had a major technological breakthrough."
Meaning: We are going back to the drawing board.

Phrase 8: "Customer satisfaction is believed assured."
Meaning: We were so far behind schedule that the customer was happy to get anything at all from us.

Phrase 9: "Preliminary operational tests were inconclusive."
Meaning: The poor thing blew up when we first tested it.

Phrase 10: "Test results were extremely gratifying."
Meaning: It works. Boy, are we surprised.

Phrase 11: "The entire concept will have to be abandoned."
Meaning: The only guy who understood the thing quit last week.

Phrase 12: "We will get back to you soon."
Meaning: You will never hear from us again.

Phrase 13: "Modifications are underway to correct certain minor difficulties."
Meaning: We threw the whole thing out and we are starting from scratch.

Phrase 14: "We have completed an extensive review of your report."
Meaning: We have read the title page of your report.

Phrase 15: "The drawings are in the mail."
Meaning: We are currently advertising to recruit someone to work on the designs.

Phrase 16: "Your point of contact is currently on out-of-town assignment."
Meaning: The person you spoke with last week is no longer with the company.

THE SEVEN "UPS" OF PROJECT MANAGEMENT

The motivational environment advocated by Triple C can be summed in some inspirational suggestions as presented below.

1. Come Up to speed with Triple C implementation process; it is essential for communication.
2. Look Up to management for direction, except when management is not Up to the task.
3. Set Up a project management office (PMO); it is needed for project coordination.
4. Reach Up for a higher project management goal; it is good motivation for the next project.
5. Lift Up the spirits of those around you; it is good for cooperation.
6. Stand Up for what is right; it is required to defend project principles.
7. Shut Up when silence is the best strategy; it is the best way to learn.

COMMON LAWS OF PROJECT MANAGEMENT

In folklore, adages and proverbs are developed to constrain the behavior of the society. In many cases, the adages represent facts that if presented as facts may not be embraced by the general population. If, however, they are presented as guiding principles to ward off "evil," then they are more apt to be believed and followed.

In modern times, adages are presented, sometimes sarcastically, as laws and principles. Some common ones are presented below. The author originated the first one in the late 1980s as a mentoring guide to his graduate students and colleagues.

Badiru's Law: "Grass is always greener where you most need it to be dead."

Parkinson's Law: "Work expands to fill the time available."

Peter's Principle: "People rise to the level of their incompetence."

Murphy's Law: "Whatever can go wrong will go wrong."

Nothing typifies Peter's Principle more that what President Warren G. Harding was quoted to have said ....

"I am not fit for this office and never should have been here."

Warren G. Harding (1865–1923), 29th president of the United States.

Such self-effacing comments or thoughts, whether expressed publicly or held privately, are common among those who have risen to the heights of leadership. Whether such ascension is due to acts of omission, commission, or by default, it is often lonely at the top, and a leader must constantly work hard to develop or acquire the wherewithal needed for the "high" office. A key part of project leadership where-withal is the ability to effectively use communication, cooperation, and coordination techniques to achieve project goals.

PARODY OF A MEETING

A meeting is not work. Meetings should work to facilitate work.

PROJECT MISSION AND VISION DICHOTOMY

Project *vision* without *action* is a *daydream.*
Project *action* without *vision* is a *nightmare.*

Project *mission* without *direction* is a *fantasy.*
Project *direction* without *mission* is a *fiction.*

IMMUTABLE LAWS OF PROJECT MANAGEMENT

Law 1: No major project is ever completed on time, within budget, with the same staff that started it, nor does the project do what it is supposed to do. It is highly unlikely that yours will be the first.

    Corollary 1: The benefits will be smaller than initially estimated, if estimates were made at all.

    Corollary 2: The system finally installed will be completed late and will not do what it is supposed to do.

    Corollary 3: It will cost more but will be technically successful.

Law 2: One advantage of fuzzy project objectives is that they let you avoid embarrassment in estimating the corresponding costs.

Law 3: The effort required to correct a project that is off course increases geometrically with time.

    Corollary 1: The longer you wait the harder it gets.

    Corollary 2: If you wait until the project is completed, it's too late.

    Corollary 3: Do it now regardless of the embarrassment.

Law 4: The project purpose statement you wrote and understand will be seen differently by everyone else.

> **Corollary 1:** If you explain the purpose so clearly that no one could possibly misunderstand, someone will.
>
> **Corollary 2:** If you do something that you are sure will meet everyone's approval, someone will not like it.

Law 5: Measurable benefits are real. Intangible benefits are not measurable; thus intangible benefits are not real.

> **Corollary 1:** Intangible benefits are real if you can prove that they are real.

Law 6: Anyone who can work effectively on a project part-time certainly does not have enough to do now.

> **Corollary 1:** If a boss will not give a worker a full-time job, you shouldn't either.
>
> **Corollary 2:** If the project participant has a time conflict, the work given by the full-time boss will not suffer.

Law 7: The greater the project's technical complexity, the less you need a technician to manage it.

> **Corollary 1:** Get the best manager you can. The manager will get the technicians.
>
> **Corollary 2:** The reverse of corollary 1 is almost never true.

Law 8: A carelessly planned project will take three times longer to complete than expected. A carefully planned project will only take twice as long.

> **Corollary 1:** If nothing can possibly go wrong, it will anyway.

Law 9: When the project is going well, something will go wrong.

> **Corollary 1:** When things cannot get any worse, they will.
>
> **Corollary 2:** When things appear to be going better, you have overlooked something.

Law 10: Project teams detest weekly progress reporting because it so vividly manifests their lack of progress.

Law 11: Projects progress rapidly until they are 90 percent complete. Then they remain 90 percent complete forever.

Law 12: If project content is allowed to change freely, the rate of change will exceed the rate of progress.

Law 13: If the user does not believe in the system, a parallel system will be developed. Neither system will work very well.

Law 14: Benefits achieved are a function of the thoroughness of the post-audit check.

> **Corollary 1:** The prospect of an independent post-audit provides the project team with a powerful incentive to deliver a good system on schedule within budget.

Law 15: No law is immutable.

This opening chapter has presented general discussions on the importance of project management in every organization. Any organizational goal can be formulated as a project, which should be managed with the usual techniques and tools of project management. The chapters that follow present various aspects of the Triple C model of project management within the context of normal business operations.

REFERENCE

Harvey, Jerry B. (1988) *The Abilene Paradox and Other Meditations on Management.* Jossey-Bass, Wiley Imprint, San Francisco, CA.

2 The Triple C Model

Tell me, and I forget;
Show me, and I remember;
Involve me, and I understand.

Chinese Proverb

We can get more work out of people through persuasion, leading to cooperation, rather than through persecution. This chapter introduces the Triple C insignia, shown in Figure 2.1, which symbolizes the integrated stages of communication, cooperation, and coordination in a project environment. As the Chinese Proverb above points out, involvement of every team member is critical for overall success of a project. The Triple C model facilitates better understanding and involvement based on foundational communication. The Triple C approach elucidates the integrated involvement of communication, cooperation, and coordination. Communication is the foundation for cooperation, which in turn is the foundation for coordination. Communication leads to cooperation, which leads to coordination, which leads to project harmony, which leads to project success.

The primary lesson of the Triple C model is not to take cooperation for granted. It must be pursued, solicited, and secured explicitly. The process of securing cooperation requires structured communication upfront. It is only after cooperation is in effect that all project efforts can be coordinated.

The Triple C model has been used effectively in practice to enhance project performance because most project problems can be traced to initial communication problems. Chapters 7 and 8 present a collection of case examples of the application of Triple C to real-life organizational problems. The Triple C approach works because it is very simple; simple to understand and simple to implement. The simplicity comes from the fact that most of the required elements of the approach are already being done within every organization, albeit in a nonstructured manner. The Triple C model puts the existing processes into a structural approach to communication, cooperation, and coordination.

ORIGIN OF THE MODEL

The idea for the Triple C model originated from a complex facility redesign project (Badiru et al., 1993) conducted for Tinker Air Force Base (TAFB) in Oklahoma City by the School of Industrial Engineering, University of Oklahoma, from 1985 through 1989. The project was a part of a reconstruction project following a disastrous fire that occurred in the base's repair/production facility in November 1984. The urgency, complexity, scope ambiguity, confusion, and disjointed directions that existed in the early days of the reconstruction effort led to the need to develop a

FIGURE 2.1   Insignia of the Triple C model.

structured approach to communication, cooperation, and coordination of the various work elements. In spite of the high pressure timing of the project, the author called a time-out-of-time (TOOT) so that a process could be developed for project communication leading to personnel cooperation, and eventually facilitating task coordination. The investment of TOOT resulted in a remarkable resurgence of cooperation where none existed at the beginning of the project. Encouraged by the intrinsic occurrence of cooperation, the process was further enhanced and formalized as the Triple C approach to the project's success. The approach was credited with the overall success of the project. The qualitative approach of Triple C complemented the technical approaches used on the project to facilitate harmonious execution of tasks. Many projects fail when the stakeholders get too wrapped up in the technical requirements at the expense of qualitative requirements. The importance of Triple C can be summarized by the diagrammatic relationship below:

Triple C → Communication → Cooperation → Coordination → Project Success

Other elements of "C," such as collaboration, commitment, and correlation, are embedded in the Triple C structure. Of course, the constraints of time, cost, and performance must be overcome all along the way.

The Triple C approach incorporates the qualitative (human) aspects of a project into overall project requirements.

INTRODUCTION TO TRIPLE C

The Triple C model was first used in 1985 and subsequently introduced in print in 1987 (Badiru, 1987). The project scenario that led to the development of the Triple C model was later documented in Badiru et al. (1993). The model is an effective project planning and control tool. The model states that project management can be enhanced by implementing it within the integrated functions summarized below:

- Communication
- Cooperation
- Coordination

The model facilitates a systematic approach to project planning, organizing, scheduling, and control. The Triple C model is distinguished from the 3C approach commonly used in military operations. The military approach emphasizes personnel management in the hierarchy of command, control, and communication. This

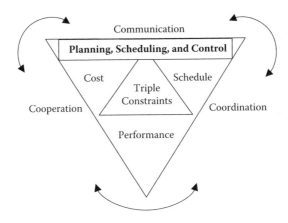

FIGURE 2.2   Triple C for planning, scheduling, and control.

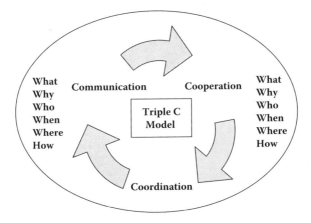

FIGURE 2.3   Triple C for who, what, why, when, where, and how.

places communication as the last function. The Triple C, by contrast, suggests communication as the first and foremost function. The Triple C model can be implemented for project planning, scheduling, and control purposes.

Figure 2.2 shows the application of Triple C for project planning, scheduling, and control within the confines of the triple constraints of cost, schedule, and performance. Each of these three primary functions of project management requires effective communication, sustainable cooperation, and adaptive coordination.

Figure 2.3 illustrates how the basic questions of who, what, why, when, where, and how revolve around the Triple C model. It highlights what must be done and when. It can also help to identify the resources (personnel, equipment, facilities, etc.) required for each effort. It points out important questions such as:

- Does each project participant know what the objective is?
- Does each participant know his or her role in achieving the objective?
- What obstacles may prevent a participant from playing his or her role effectively?

Triple C can mitigate disparity between idea and practice because it explicitly solicits information about the critical aspects of a project in terms of the following queries:

- Who
- What
- Why
- When
- Where
- How

TYPES OF COMMUNICATION

- Verbal
- Written
- Body language
- Visual tools (e.g., graphical tools)
- Sensual (use of all five senses: sight, smell, touch, taste, hearing)
- Simplex (unidirectional)
- Half-duplex (bidirectional with time lag)
- Full-duplex (real-time dialogue)
- One-on-one
- One-to-many
- Many-to-one

TYPES OF COOPERATION

- Proximity
- Functional
- Professional
- Social
- Romantic
- Power influence
- Authority influence
- Hierarchical
- Lateral
- Cooperation by intimidation
- Cooperation by enticement

TYPES OF COORDINATION

- Teaming
- Delegation
- Supervision
- Partnership
- Token-passing
- Baton hand-off

TRIPLE C QUESTIONS

Questioning is the best approach to getting information for effective project management. Everything should be questioned. By upfront questions, we can preempt and avert project problems later on. Typical questions to ask under Triple C approach are the following:

- What is the purpose of the project?
- Who is in charge of the project?
- Why is the project needed?
- Where is the project located?
- When will the project be carried out?
- How will the project contribute to increased opportunities for the organization?
- What is the project designed to achieve?
- How will the project affect different groups of people within the organization?
- What will be the project approach or methodology?
- What other groups or organizations will be involved (if any)?
- What will happen at the end of the project?
- How will the project be tracked, monitored, evaluated, and reported?
- What resources are required?
- What are the associated costs of the required resources?
- How do the project objectives fit the goal of the organization?
- What respective contribution is expected from each participant?
- What level of cooperation is expected from each group?
- Where is the coordinating point for the project?

The key to getting everyone on board with a project is to ensure that task objectives are clear and comply with the principle of SMART as outlined below and expanded later in Chapter 3:

Specific: Task objective must be specific.
Measurable: Task objective must be measurable.
Aligned: Task objective must be achievable and aligned with overall project goal.
Realistic: Task objective must be realistic and relevant to the organization.
Timed: Task objective must have a time basis.

If a task has the above intrinsic characteristics, then the function of communicating the task will more likely lead to personnel cooperation.

COMMUNICATION

Communication makes working together possible. The communication function of project management involves making all those concerned become aware of project requirements and progress. Those who will be affected by the project directly or indirectly, as direct participants or as beneficiaries, should be informed of the following as appropriate:

- Scope of the project
- Personnel contribution required
- Expected cost and merits of the project
- Project organization and implementation plan
- Potential adverse effects if the project should fail
- Alternatives, if any, for achieving the project goal
- Potential direct and indirect benefits of the project

The communication channel must be kept open throughout the project life cycle. In addition to internal communication, appropriate external sources should also be consulted. The project manager must do the following:

- Exude commitment to the project
- Utilize the communication responsibility matrix
- Facilitate multichannel communication interfaces
- Identify internal and external communication needs
- Resolve organizational and communication hierarchies
- Encourage both formal and informal communication links

Readers should refer to Chapter 3 for additional exposition of the above guidelines. When clear communication is maintained between management and employees and among peers, many project problems can be averted. Project communication may be carried out in one or more of the following formats:

- One-to-many
- One-to-one
- Many-to-one
- Written and formal
- Written and informal
- Oral and formal
- Oral and informal
- Nonverbal gestures

Good communication is affected when what is implied is perceived as intended. Effective communications are vital to the success of any project. Despite the awareness that proper communication forms the blueprint for project success, many organizations still fail in their communication functions. The study of communication is complex. Factors that influence the effectiveness of communication within a project organization structure include the following.

1. **Personal perception.** Each person perceives events on the basis of personal psychological, social, cultural, and experimental background. As a result, no two people can interpret a given event the same way. The nature

of events is not always the critical aspect of a problem situation. Rather, the problem is often the different perceptions of the different people involved.

2. **Psychological profile.** The psychological makeup of each person determines personal reactions to events or words. Thus, individual needs and level of thinking will dictate how a message is interpreted.

3. **Social Environment.** Communication problems sometimes arise because people have been conditioned by their prevailing social environment to interpret certain things in unique ways. Vocabulary, idioms, organizational status, social stereotypes, and economic situation are among the social factors that can thwart effective communication.

4. **Cultural background.** Cultural differences are among the most pervasive barriers to project communications, especially in today's multinational organizations. Language and cultural idiosyncrasies often determine how communication is approached and interpreted.

5. **Semantic and syntactic factors.** Semantic and syntactic barriers to communications usually occur in written documents. Semantic factors are those that relate to the intrinsic knowledge of the subject of the communication. Syntactic factors are those that relate to the form in which the communication is presented. The problems created by these factors become acute in situations where response, feedback, or reaction to the communication cannot be observed.

6. **Organizational structure.** Frequently, the organization structure in which a project is conducted has a direct influence on the flow of information and, consequently, on the effectiveness of communication. Organization hierarchy may determine how different personnel levels perceive a given communication.

7. **Communication media.** The method of transmitting a message may also affect the value ascribed to the message and, consequently, how it is interpreted or used. The common barriers to project communications are:

- Inattentiveness
- Lack of organization
- Outstanding grudges
- Preconceived notions
- Ambiguous presentation
- Emotions and sentiments
- Lack of communication feedback
- Sloppy and unprofessional presentation
- Lack of confidence in the communicator
- Lack of confidence by the communicator
- Low credibility of the communicator
- Unnecessary technical jargon
- Too many people involved
- Untimely communication
- Arrogance or imposition
- Lack of focus

Some suggestions on improving the effectiveness of communication are presented next. The recommendations may be implemented as appropriate for any of the forms of communications listed earlier. The recommendations are for both the communicator and the audience.

1. Never assume that the integrity of the information sent will be preserved as the information passes through several communication channels. Information is generally filtered, condensed, or expanded by the receivers before relaying it to the next destination. When preparing a communication that needs to pass through several organization structures, one safeguard is to compose the original information in a concise form to minimize the need for recomposition of the project structure.

2. Give the audience a central role in the discussion. A leading role can help make a person feel a part of the project effort and responsible for the projects' success. He or she can then have a more constructive view of project communication.

3. Do homework and think through the intended accomplishment of the communication. This helps eliminate trivial and inconsequential communication efforts.

4. Carefully plan the organization of the ideas embodied in the communication. Use indexing or points of reference whenever possible. Grouping ideas into related chunks of information can be particularly effective. Present the short messages first. Short messages help create focus, maintain interest, and prepare the mind for the longer messages to follow.

5. Highlight why the communication is of interest and how it is intended to be used. Full attention should be given to the content of the message with regard to the prevailing project situation.

6. Elicit the support of those around you by integrating their ideas into the communication. The more people feel they have contributed to the issue, the more expeditious they are in soliciting the cooperation of others. The effect of the multiplicative rule can quickly garner support for the communication purpose.

7. Be responsive to the feelings of others. It takes two to communicate. Anticipate and appreciate the reactions of members of the audience. Recognize their operational circumstances and present your message in a form they can relate to.

8. Accept constructive criticism. Nobody is infallible. Use criticism as a springboard to higher communication performance.

9. Exhibit interest in the issue in order to arouse the interest of your audience. Avoid delivering your messages as a matter of a routine organizational requirement.

10. Obtain and furnish feedback promptly. Clarify vague points with examples.

11. Communicate at the appropriate time, at the right place, to the right people.

12. Reinforce words with positive action. Never promise what cannot be delivered. Value your credibility.

13. Maintain eye contact in oral communication and read the facial expressions of your audience to obtain real-time feedback.

14. Concentrate on listening as much as speaking. Evaluate both the implicit and explicit meanings of statements.
15. Document communication transactions for future references.
16. Avoid asking questions that can be answered yes or no. Use relevant questions to focus the attention of the audience. Use questions that make people reflect upon their words, such as, "How do you think this will work?" compared to "Do you this will work?"
17. Avoid patronizing the audience. Respect their judgment and knowledge.
18. Speak and write in a controlled tempo. Avoid emotionally charged voice inflections.
19. Create an atmosphere for formal and informal exchange of ideas.
20. Summarize the objectives of the communication and how they will be achieved.

Figure 2.4 shows an example of a design of a communication responsibility matrix. A communication responsibility matrix shows the linking of sources of communication and targets of communication. Cells within the matrix indicate the subject of the desired communication. There should be at least one filled cell in each row and each column of the matrix. This ensures that each individual of a department has at least one communication source or target associated with him or her. With a communication responsibility matrix, a clear understanding of what needs to be communicated to whom can be developed. Communication in a project environment can take any of several forms. The specific needs of a project may dictate the most appropriate

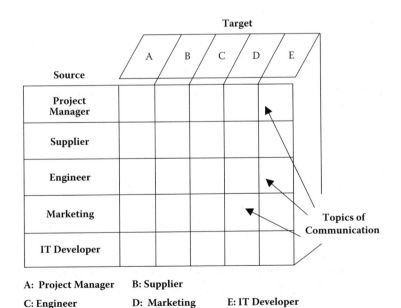

A: Project Manager   B: Supplier

C: Engineer   D: Marketing   E: IT Developer

FIGURE 2.4   Triple C communication matrix.

mode. Three popular computer communication modes are discussed next in the context of communicating data and information for project management.

Simplex communication. This is a unidirectional communication arrangement in which one project entity initiates communication to another entity or individual within the project environment. The entity addressed in the communication does not have the mechanism or capability for responding to the communication. An extreme example of this is a one-way, top-down communication from top management to the project personnel. In this case, the personnel have no communication access or input to top management. A budget-related example is a case where top management allocates budget to a project without requesting and reviewing the actual needs of the project. Simplex communication is common in authoritarian organizations.

Half-duplex communication. This is a bidirectional communication arrangement whereby one project entity can communicate with another entity and receive a response within a certain time lag. Both entities can communicate with each other, but not at the same time. An example of half-duplex communication is a project organization that permits communication with top management without a direct meeting. Each communicator must wait for a response from the target of the communication. Request and allocation without a budget meeting is another example of half-duplex data communication in project management.

Full-duplex communication. This involves a communication arrangement that permits a dialogue between the communicating entities. Both individuals and entities can communicate with each other at the same time or face-to-face. As long as there is no clash of words, this appears to be the most receptive communication mode. It allows participative project planning in which each project personnel has an opportunity to contribute to the planning process.

Figure 2.5 presents a graphical representation of the communication modes discussed above. Each member of a project team needs to recognize the nature of the prevailing communication mode in the project. Management must evaluate the prevailing communication structure and attempt to modify it if necessary to enhance project functions. An evaluation of who is to communicate with whom about what may help improve the project data/information communication process. A communication matrix may include notations about the desired modes of communication between individuals and groups in the project environment.

COOPERATION

The cooperation of the project personnel must be explicitly elicited. Merely voicing consent for a project is not enough assurance of full cooperation. The participants and beneficiaries of the project must be convinced of the merits of the project. Some of the factors that influence cooperation in a project environment include personnel requirements, resource requirements, budget limitations, past experiences, conflicting priorities, and lack of uniform organizational support. A structured approach to seeking cooperation should clarify the following:

- Cooperative efforts required
- Precedents for future projects

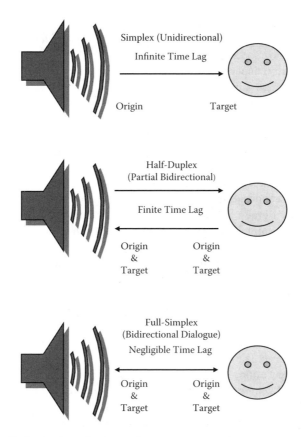

FIGURE 2.5   Project communication modes.

- Implication of lack of cooperation
- Criticality of cooperation to project success
- Organizational impact of cooperation
- Time frame involved in the project
- Rewards of good cooperation

Cooperation is a basic virtue of human interaction. More projects fail due to a lack of cooperation and commitment than any other project factors. To secure and retain the cooperation of project participants, you must elicit a positive first reaction to the project. The most positive aspects of a project should be the first items of project communication. For project management, there are different types of cooperation that should be understood.

Functional cooperation. This is cooperation induced by the nature of the functional relationship between two groups. The two groups may be required to perform related functions that can only be accomplished through mutual cooperation.

Social cooperation. This is the type of cooperation effected by the social relationship between two groups. The prevailing social relationship motivates cooperation that may be useful in getting project work done.

Legal cooperation. Legal cooperation is the type of cooperation that is imposed through some authoritative requirement. In this case, the participants may have no choice other than to cooperate.

Administrative cooperation. This is cooperation brought on by administrative requirements that make it imperative that two groups work together on a common goal.

Associative cooperation. This type of cooperation may also be referred to as collegiality. The level of cooperation is determined by the association that exists between two groups.

Proximity cooperation. Cooperation due to the fact that two groups are geographically close is referred to as proximity cooperation. Being close makes it imperative that the two groups work together.

Dependency cooperation. This is cooperation caused by the fact that one group depends on another group for some important aspect. Such dependency is usually of a mutual two-way nature. One group depends on the other for one thing while the latter group depends on the former for some other thing.

Imposed cooperation. In this type of cooperation, external agents must be employed to induce cooperation between two groups. This is applicable for cases where the two groups have no natural reason to cooperate. This is where the approaches presented earlier for seeking cooperation can become very useful.

Lateral cooperation. Lateral cooperation involves cooperation with peers and immediate associates. Lateral cooperation is often easy to achieve because existing lateral relationships create an environment that is conducive for project cooperation.

Vertical cooperation. Vertical or hierarchical cooperation refers to cooperation that is implied by the hierarchical structure of the project. For example, subordinates are expected to cooperate with their vertical superiors.

Whichever type of cooperation is available in a project environment, the cooperative forces should be channeled toward achieving project goals. Documentation of the prevailing level of cooperation is useful for winning further support for a project. Clarification of project priorities will facilitate personnel cooperation. Relative priorities of multiple projects should be specified so that they are a priority to all groups within the organization. Some guidelines for securing cooperation for most projects the following:

- Establish achievable goals for the project.
- Clearly outline the individual commitments required.
- Integrate project priorities with existing priorities.
- Eliminate the fear of job loss due to industrialization.
- Anticipate and eliminate potential sources of conflict.
- Use an open-door policy to address project grievances.
- Remove skepticism by documenting the merits of the project.

Commitment. Cooperation must be supported with commitment. To cooperate is to support the ideas of a project. To commit is to willingly and actively participate in project efforts again and again through the thick and thin of the project. Provision of resources is one way that management can express commitment to a project.

Triple C + Commitment = Project Success

TABLE 2.1

Example of Responsibility Matrix for Project Coordination

Tasks	Person Responsible				Status of Task			
	Staff A	Staff B	Staff C	Mgr	31 Jan	15 Feb	28 Mar	21 Apr
Brainstorming meeting	R	R	R	R	D			
Identify speakers				R		O		
Select seminar location	I	R	R			O		
Select banquet location	R	R				D		
Prepare publicity materials		C	R	I	O	O	D	
Draft brochures		C	R					D
Develop schedule			R			L	L	
Arrange for visual aids			R		L	L	L	
Coordinate activities			R				L	
Periodic review of tasks	R	R	R	S				D
Monitor progress of program	C	R	R			O	L	
Review program progress	R				O	O	L	L
Closing arrangements	R							L
Postprogram review and evaluation	R	R	R	R			D	

Note: Responsibility codes, R = Responsible; I = Inform; S = Support; C = Consult.
Task Codes, D = Done; O = On Track; L = Late.

COORDINATION

After the communication and cooperation functions have successfully been initiated, the efforts of the project personnel must be coordinated. Coordination facilitates harmonious organization of project efforts. The construction of a responsibility chart can be very helpful at this stage. A responsibility chart is a matrix consisting of columns of individual or functional departments and rows of required actions. Cells within the matrix are filled with relationship codes that indicate who is responsible for what. Table 2.1 illustrates an example of a responsibility matrix for the planning for a seminar program. The matrix helps avoid neglecting crucial communication requirements and obligations. It can also help resolve the following questions:

- Who is to do what?
- How long will it take?
- Who is to inform whom of what?
- Whose approval is needed for what?
- Who is responsible for which results?
- What personnel interfaces are required?
- What support is needed from whom and when?

CONFLICT RESOLUTION USING TRIPLE C APPROACH

Conflicts can and do develop in any work environment. Conflicts, whether intended or inadvertent, prevent an organization from getting the most out of the work force. When implemented as an integrated process, the Triple C model can help avoid conflicts in a project. When conflicts do develop, it can help in resolving the conflicts. The key to conflict resolution is open and direct communication, mutual cooperation, and sustainable coordination. Several sources of conflicts can exist in a project. Some of these are discussed below.

Schedule conflict. Conflicts can develop because of improper timing or sequencing of project tasks. This is particularly common in large multiple projects. Procrastination can lead to having too much to do at once, thereby creating a clash of project functions and discord among project team members. Inaccurate estimates of time requirements may lead to infeasible activity schedules. Project coordination can help avoid schedule conflicts.

Cost conflict. Project cost may not be generally acceptable to the clients of a project. This will lead to project conflict. Even if the initial cost of the project is acceptable, a lack of cost control during implementation can lead to conflicts. Poor budget allocation approaches and the lack of a financial feasibility study will cause cost conflicts later on in a project. Communication and coordination can help prevent most of the adverse effects of cost conflicts.

Performance conflict. If clear performance requirements are not established, performance conflicts will develop. Lack of clearly defined performance standards can lead each person to evaluate his or her own performance based on personal value judgments. In order to uniformly evaluate quality of work and monitor project progress, performance standards should be established by using the Triple C approach.

Management conflict. There must be a two-way alliance between management and the project team. The views of management should be understood by the team. The views of the team should be appreciated by management. If this does not happen, management conflicts will develop. A lack of a two-way interaction can lead to strikes and industrial actions, which can be detrimental to project objectives. The Triple C approach can help create a conducive dialogue environment between management and the project team.

Technical conflict. If the technical basis of a project is not sound, technical conflict will develop. New industrial projects are particularly prone to technical conflicts because of their significant dependence on technology. Lack of a comprehensive technical feasibility study will lead to technical conflicts. Performance requirements and systems specifications can be integrated through the Triple C approach to avoid technical conflicts.

Priority conflict. Priority conflicts can develop if project objectives are not defined properly and applied uniformly across a project. Lack of a direct project definition can lead each project member to define his or her own goals, which may be in conflict with the intended goal of a project. Lack of consistency of the project mission is another potential source of priority conflicts. Overassignment of responsibilities with no guidelines for relative significance levels can also lead to priority conflicts. Communication can help defuse priority conflicts.

Resource conflict. Resource allocation problems are a major source of conflict in project management. Competition for resources, including personnel, tools, hardware, software, and so on, can lead to disruptive clashes among project members. The Triple C approach can help secure resource cooperation.

Power conflict. Project politics lead to a power play, which can adversely affect the progress of a project. Project authority and project power should be clearly delineated. Project authority is the control that a person has by virtue of his or her functional post. Project power relates to the clout and influence that a person can exercise as a result of connections within the administrative structure. People with popular personalities can often wield a lot of project power in spite of low or nonexistent project authority. The Triple C model can facilitate a positive marriage of project authority and power to the benefit of project goals. This will help define clear leadership for a project.

Personality conflict. Personality conflict is a common problem in projects involving a large group of people. The larger the project, the larger the size of the management team needed to keep things running. Unfortunately, the larger management team creates an opportunity for personality conflicts. Communication and cooperation can help defuse personality conflicts. In summary, conflict resolution through Triple C can be achieved by observing the following guidelines:

1. Confront the conflict and identify the underlying causes.
2. Be cooperative and receptive to negotiation as a mechanism for resolving conflicts.
3. Distinguish between proactive, inactive, and reactive behaviors in a conflict situation.
4. Use communication to defuse internal strife and competition.
5. Recognize that short-term compromise can lead to long-term gains.
6. Use coordination to work toward a unified goal.
7. Use communication and cooperation to turn a competitor into a collaborator.

It is the little and often neglected aspects of a project that lead to project failures. Several factors may constrain the project implementation. All the relevant factors can be evaluated under the Triple C model right from the project-initiation stage. The adaptation of the nursery rhyme below illustrates the importance of the little "things" in a project:

FALL OF THE KINGDOM

For the want of a nail, the horse shoe was lost;

For the want of the horse shoe, the horse was lost;

For the want of the horse, the rider was lost;

For the want of the rider, the message was lost;

For the want of the message, the battle was lost;

For the want of the battle, the war was lost;

For the want of the war, the Kingdom was lost.

All for the want of a nail!

FAILURE OF THE PROJECT

For the want of communication, the data was lost;

For the lack of the data, the information was lost;

For the lack of the information, the decision was lost;

For the lack of the decision, the cooperation was lost;

For the lack of the cooperation, the planning was lost;

For the lack of the planning, the coordination was lost;

For the lack of the coordination, the project failed.

All for the want of simple communication!

APPLICATION OF TRIPLE C TO STEPS

Having now understood the intrinsic elements of Triple C, we can see how and where it could be applicable to the steps of project management. Communication explains project scope and requirements through the stages of planning, organizing, scheduling, and control. Cooperation is required to get human resource buy-in and stakeholder endorsement across all facets of planning, organizing, scheduling, and control. Coordination facilitates adaptive interfaces over all the elements of planning, organizing, scheduling, and control. The Triple C model should be implemented as an iterative loop process that moves a project through the communication, cooperation, and coordination functions as suggested in Figure 2.6.

COMMUNICATION
COOPERATION
COORDINATION

FIGURE 2.6   Triple C iterative loop process.

DMAIC AND TRIPLE C

Many organizations now explore Six Sigma define, measure, analyze, improve, and control (DMAIC) methodology and associated tools to achieve better project performance. Six sigma means six standard deviations from a statistical performance average. The six sigma approach allows for no more than 3.4 defects per million parts in manufactured goods or 3.4 mistakes per million activities in a service operation. To explain the effect of the six sigma approach, consider a process that is 99% perfect. That process will produce 10,000 defects per million parts. With six sigma, the process will need to be 99.99966% perfect in order to produce only 3.4 defects per million. Thus, Six Sigma is an approach that moves a process toward perfection. Six Sigma, in effect, reduces variability among products produced by the same process. By contrast, the lean approach is designed to reduce/eliminate waste in the production process.

Six Sigma provides a roadmap for the five major steps of DMAIC (define, measure, analyze, improve, and control), which are applicable to the planning and control steps of project management. We cannot improve what we cannot measure. Triple C provides a sustainable approach to obtaining cooperation and coordination for DMAIC during improvement efforts. DMAIC requires project documentation and reporting, which coincide with project control requirements.

This chapter has presented a general introduction to the Triple C approach. The next three chapters are devoted to discussions of communication, cooperation, and coordination. A summary of lessons to be inferred from a Triple C approach are the following:

- Use proactive planning to initiate project functions.
- Use preemptive planning to avoid project pitfalls.
- Use meetings strategically. Meeting is not *work*. Meeting should be done to facilitate work.
- Use project assessment to properly frame the problem, adequately define the requirements, continually ask the right questions, cautiously analyze risks, and effectively scope the project.
- Be bold to terminate a project when termination is the right course of action. Every project needs an exit plan. In some cases, there is victory in capitulation.

An implementation template for the Triple C model is presented in Chapter 6. The sustainability of the Triple C approach is summarized below:

1. For effective communication, create good communication channels.
2. For enduring cooperation, establish partnership arrangements.
3. For steady coordination, use a workable organization structure.

REFERENCES

Badiru, Adedeji B., B. L. Foote, L. Leemis, A. Ravindran, and L. Williams. (1993) "Recovering from a Crisis at Tinker Air Force Base," PM Network, Vol. 7, No. 2, Feb. 1993, pp. 10–23.

Badiru, Adedeji B., (1987) "Communication, Cooperation, Coordination: The Triple C of Project Management," in Proceedings of 1987 IIE Spring Conference, Washington, DC, May 1987, pp. 401–404.

3 Project Communication

No man can efficiently direct work about which he knows nothing.

Col. Thurman H. Bane, Air School of Application, USA, 1919

Communication is the basis of everything and is thus the key to effective project management. Even in biblical times, the importance of project communication was contained in the chronicle of the Tower of Babel, whereby it was reported that God caused a construction project to fail by interrupting communication through the creation of multiple languages.

Without a common basis for communication, any project is bound to fail. Communication is the primary basis for acquiring knowledge about the task at hand. Communication is the basis for project performance in any organization. Information is power, and those who have it will hold the key to project success. Information is transmitted through effective communication. In the present information-drive society, having access to project information paves the way for every individual's contribution. The Triple C approach to project management puts communication first and foremost in project endeavors.

While extensive communication is required for project implementation, care must be taken to avoid being dismissive of the negative aspects of a project. "Accentuate the positive, eliminate the negative" is a popular cliché. But under Triple C, a more realistic approach suggests an alternate cliché that says "accentuate the positive, mitigate the negative." To claim to eliminate all negatives is to engage in impractical denial. Questions must be raised about any project right from the beginning. The communication tools of Triple C can then be used to resolve the questions that arise. Typical early questions should address the what, who, why, where, how, and when of a project. The Triple C questions from Chapter 2 are repeated below for ease of reference:

- What is the purpose of the project?
- Who is in charge of the project?
- Why is the project needed?
- Where is the project located?
- When will the project be carried out?
- How will the project contribute to increased opportunities for the organization?
- What is the project designed to achieve?
- How will the project affect different groups of people within the organization?

- What will be the project approach or methodology?
- What other groups or organizations will be involved (if any)?
- What will happen at the end of the project?
- How will the project be tracked, monitored, evaluated, and reported?
- What resources are required?
- What are the associated costs of the required resources?
- How do the project objectives fit the goal of the organization?
- What respective contribution is expected from each participant?
- What level of cooperation is expected from each group?
- Where is the coordinating point for the project?

Communication must be direct and structured. Epileptic communication, using an on-off style of personnel interaction, does not provide an avenue for project comprehension, consistency, and team cohesion. It is essential to double-check and even triple-check lines of communication for fidelity before proceeding too deep into the project scope. Communication must be done in a timely manner and be concise. Prompt communication increases the chances of being received accurately. If communication degenerates to protracted pontification, then it becomes useless. If communication elongates to the point of being "timed out," then it becomes "tuned out." As a general guide, project communicators should understand the concept of first-in-first-out understood (FIFU) as well as first-on-first-understood (FOFU). This is important because during the transient state of communication, project requirements are likely to be dynamic and volatile. Expediency is essential to getting the contents of communication across effectively. Use targeted communication strategies. Use different levels of communication details for different levels of responsibility.

OPERATIONAL IMPORTANCE OF PROJECT COMMUNICATION

The foundational importance of communication in project management was emphasized in an online quick quiz note by the Project Management Institute (PMI). The service is an online quiz-and-answer or ask-the-expert membership community service of PMI. The quiz-and-answer is presented here:

THE QUIZ (FROM *PMI COMMUNITY POST*, SEPTEMBER 14, 2007):

I know that project managers may spend up to 90 percent of their time in communication, but that seems like a lot. How do I spend that much time in a meaningful way?

THE ANSWER (PROVIDED BY BARBEE DAVIS, M.A., PHR, PMP):

It depends on your project management experience and that of your team. Great communication can make or break a project, so to devote extensive time in this Knowledge Area is reasonable. The challenge is to be sure to use that time for activities that bring value to your project.

It is a frequent misconception that most time spent in communication should be focused on your team. In fact, too much communication can be perceived as over-supervision of project work, which can be detrimental and de-motivate your team. Also, as your experience increases, you become more efficient so ordinary project management tasks take fewer hours. And as team members gain experience, they typically become more self-sufficient.

When both you are your team are working efficiently, you're ready to move up the maturity scale and redirect those "extra" chunks of time previously needed for team communication. Here are some suggestions:

- Free yourself for a more strategic project management role. Train a team member to perform some of your routine tasks or advocate for an automated software system.
- Take time to find out the business strategy goals underlying this project. The success of this project may be more strategic than just delivery of a product or service.
- Develop a system to capture lessons learned during the project. When you wait until the end, it's easy to forget the problems, why they occurred and the reasons you solved them as you did.

If you meet with customers to document requirements and normally don't see them again until the project is over, change this pattern. Show the customer partial developments along the way to avoid major scope changes late in the project.

- Invest more time to work on interpersonal team issues that can sabotage even a team with exceptional skill sets.
- Research new project management ideas that could streamline this and future projects. Check the impact of your project on projects that may be planned by your company in the future. Your current project may be more strategic than you realize. Look at the next step for your company to move up a level in project management maturity. Then champion some steps to head the organization in that direction.

Team communication is extremely important and should not be neglected. But these additional suggestions for how you might spend any extra communication time can help you spend it in a meaningful way.

COMMUNICATING ACROSS PROJECT STEPS

Communication is the lifeline of any project organization. Blocks and hierarchies contained in the project organization are linked and sustained through effective communication. Figure 3.1 shows a network of communication across project entities. Static organization structures that are implemented as de facto requirements do not serve the purpose of securing across-the-board cooperation. A project schedule represents the sequence of activities and tasks in pursuit of the end product of the project. The elements in the sequence are linked through effective baton-passing communication from one stage of the project to the next. Without each stage of recognizing and understanding the functions, goals, roles, and responsibilities within the project, the overall project outlook runs the danger of failure.

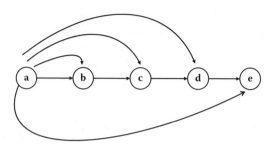

FIGURE 3.1   Triple C communication networking.

Control is really a misnomer in the context of contemporary projects. In the traditional sense, "monitoring and control" were generally accepted. But the modern work environment requires a more collaborative approach rather than a strict monitoring practice followed by an authoritarian control procedure. "Control" seems to conjure up an image of sinister intention on the part of the person doing the control. For this reason, a term that is becoming more frequently used is "project tracking and reporting." Communication should form the basis for corrective control in a modern project. In fact, proactive and preemptive communication can help to reduce the need for stringent control.

Networking is the process of building mutually beneficial relationships among project team members. There are two opposing views of networking. On one side are those who erroneously believe that networking is about getting to know people with the end result of getting something from them. On the more astute side are those who see networking as an opportunity to "give back" to those who need assistance. Networking cannot work without effective communication. Networking should be used as a way to communicate project information bottom-up and top-down. This requires a multilateral involvement approach to networking. The importance of being fully involved is seen in the Chinese proverb presented at the beginning of Chapter 2.

This implies that involvement through the Triple C approach can enhance the understanding of every member of the project. When the project requirements are fully understood, everyone would be ready to participate and perform at the highest level.

THE FIVE SENSES FOR COMMUNICATION

Communication can be effected through various means. In addition to our normal verbal communication, we can employ all five senses to get the project information across to the audience. Figure 3.2 illustrates the five senses that can be employed in the communication process. Making eye contact is a form of communication. Hearing and speaking are forms of communication. Touching is a form of communication. Smelling is a form of communication. Even tasting (e.g., at a project reception) is a form of communication. All these should be employed where applicable as a part of the Triple C communication process.

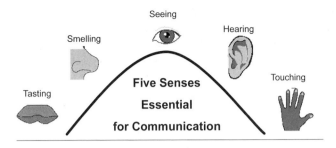

FIGURE 3.2 Five senses essential for communication.

AUTOMATIC NEGATIVE THINKERS WITHIN THE PROJECT SYSTEM

Owing to an act of nature or social distortion, some people are automatic negative thinkers (ANTs) and only see the negative aspects of everything. Effective communication is a good way to overcome such negative tendencies. Those who think negatively are perhaps conditioned to do so over years of mistrust and direct adverse experiences. They require reassurances to help them overcome those past experiences. Clear objectives and explicit outline of project requirements, expectations, and respective roles can often help to overcome, or at least mitigate, the distrust of ANTs.

COMMUNICATING CLEAR PROJECT OBJECTIVES

Project objectives provide directions to a team concerning what needs to be done, when, and how. Thus, project objectives must be communicated clearly and positively, and with enthusiasm. Objectives must be realistic and achievable. As an anecdotal example, a project manager once confided in the author by saying, "In all my years of experience, I never missed a deadline until I came to this organization." This is a classic indication of unreasonable requirements, unclear objectives, and unrealistic expectations.

SMART APPROACH TO PROJECT OBJECTIVES

A smart approach to developing and communicating project objectives uses the acronym for specific, measurable, aligned, realistic, and timed (SMART). Working hard is not the only option. Working smart provides better long-term benefits. The author often uses the analogy of a novice swimmer. The novice swimmer works the hardest, huffing and puffing, but with very little distance covered. By contrast, an expert swimmer covers more distance with very little effort. So, an observation of "hard work" does not necessarily imply work progress. The SMART notion of project objectives is expanded below:

Specific: Project objectives must be specific, explicit, and unambiguous. Objectives that are not specific are subject to misinterpretations and misuse.
Measurable: Project objectives should be designed to be measurable. Any factor that cannot be measured cannot be tracked, evaluated, or controlled.

Aligned: A project's goals and objectives must be aligned with the core strategy of an organization and relevant to prevailing needs. If not aligned, an objective will have misplaced impacts.

Realistic: A project and its essential elements must be realistic and achievable. It is good to "dream" and have lofty ideas of what can be achieved. But if those pursuits are not realistic, a project will just end up "spinning wheels" without any significant achievements.

Timed: Timing is the standardized basis for work accomplishment. If project expectations are not normalized against time, there would be no basis for accurate assessment of performance.

To further expatiate on the SMART approach, consider the following meanings. Specific means that an observable action, behavior, or achievement is described. It also means that the work links to a rate of performance, frequency, percentage, or other quantifiable measure. For some jobs, being specific can, itself, be nebulous. However, to whatever extent possible and reasonable, we should try to achieve specificity. This ensures that managers, supervisors, and employees share the same expectations.

- Example: As an objective, "To type all correspondence effectively" is not a precise description of a desired level of performance; it defines no basis for assessment. The objective would be better if it read, "Ensure correspondence is formatted correctly and complies with the department style guide for formatting, grammar, and usage as determined by managerial guidelines." This objective specifies a standard (the department style guide) by which to measure the work.
- The objective should be specific about the end result and not about the way in which it is achieved.
- The idea of objectives is to give greater responsibility for the results of the work to the employee. This means they should be able to decide how to reach the objective, but the objective itself must be very clearly defined.

The word measurable means observable or verifiable, which implies that a method or procedure must be in place to track and assess the behavior or action on which the objective focuses and the quality of the outcome. Since not all work lends itself to measurability, objectives can be written in a way that focuses on observable or verifiable behavior or results, rather than on measurable results.

Setting an objective that requires correspondence to be formatted correctly and free of typographical errors is acceptable if a system exists to measure achievement of the specified level of performance. In the case of ensuring that correspondence complies with the department style guide, a measurement system is in place (the managerial review). If no measurement system exists, the project manager must be able to monitor performance to ensure it complies with the specified objective.

An aligned objective provides a conceptual basis to draw a linkage line from the objective to other factors throughout the project.

- It means that the objectives throughout the organization pull in the same direction. In this way the performance of the project team and whole organization is improved.
- Project managers must have a clear understanding of their own objectives before they can work with project team members to establish their job objectives. This is one of the key building blocks of performance assessment in project management. If managers know the functions on which people actually are spending time, they can make meaningful improvements in organizational performance by ensuring effort is focused on work that the organization values and by eliminating inefficient processes. Job objectives align work with organizational goals and the mission, drawing the line of sight between the employee's work, the work unit's goals, the project functions, and the organization's success.

The letter "R" in SMART has two meanings that are both important: realistic and relevant.

- Realistic has two meanings:
  - The achievement of an objective is something an employee or a team can do that will support a work unit's goal. The objective should be sufficiently complex to challenge the individual or team, but not so complex that it cannot be accomplished. At the same time, it should not be so easy that it does not bring value to the individual or the team.
  - The objective should be achievable within the time and resources available to the project. In other words, the objective is designed to fit within the box of the triple constraints (Figure 2.2).
- Relevant implies it is important for the advancement of the employee and the organization.

The "relevant" element of SMART requires meaningful distinctions between employees at different levels in the hierarchy of needs, functional responsibilities, and personal capabilities.

An objective that is timed, timely, or time-bound means there is a point in time when the objective will start or when it will be completed. This is an important basis for project termination. It should be recalled that termination is one of the control aspects of project management.

- When developing objectives for ongoing work, it is essential to specify a date (month, day, or year) when the assessment period is to begin. If it is a short-term or project-related objective, we should specify when the assessment period is to be completed.
- The date component can be expressed in two ways: relatively or specifically.

- Relative uses time spans, such as "in six months." The date is relative to today.
- Specific uses hard dates, such as "on September 2nd."

While both may be used, the best practice is to use specific dates because there is a greater tendency to repeatedly push back on relative deadlines.

Good examples of explicit and clear work objectives can be found in the government job instruction manuals. A specific sample is presented below from the government acquisitions process. To communicate job objectives is one thing, but for a worker to work accordingly is another. That is where the application of "cooperation" from the Triple C model comes in. Knowing the big picture helps each person to have a sense of relevance to the project, and consequently helps to build cooperation.

COMMUNICATING PROJECT OBJECTIVES

Functional Area	Objectives	Contributing Factors
Contracting	**Target: supervisor with contracting responsibility** Accomplishes necessary actions to ensure thorough and coherent acquisition, pricing, and contract documents, which effectively implement approved strategies, meet all legal and regulatory requirements, and provide deliverables at fair and reasonable prices. Contracts and documentation must meet legal and regulatory requirements with no more than X% of the documents cited for major deficiencies through the clearance process, or other external reviews such as semiannual self-inspections. Establishes reasonable milestones for procurement completion in concert with program needs, and meets those milestones X% of the time.	Technical proficiency Critical thinking
Contracting	**Target: supervisor with contracting responsibility** Consistently exhibits the ability to synthesize information and provide sound advice, guidance, and direction in order to resolve complex contracting (business or strategic, depending on level of the position) issues. Takes timely and aggressive action to elevate problems and propose solutions to the appropriate level of management as needed. Adapts to changing priorities and program goals of the customer and proposes contracting (business, strategic) solutions to effectively implement those changes.	Critical thinking Communication Customer focus
Engineering	**Title: technical policy and direction** Work with the division chief, technical director, and branch technical advisors to implement technical policy and direction and track compliance; collaborate with division and branch technical leadership to establish new methods,	Technical proficiency Communication

COMMUNICATING PROJECT OBJECTIVES (continued)

Functional Area	Objectives	Contributing Factors
	procedures, and practices supporting technical policy and direction; assist the division leadership in establishing policy and direction consistent with division objectives and branch capabilities. For those (specs and standards) documents that the branch is identified as the "preparing activity," the incumbent will require the assigned/responsible engineer to define and implement a review/revision plan (definition of tasks, schedule, and required resources) and submit to the branch technical advisor for approval. The incumbent will provide the necessary resources to execute the plan and monitor review/revision progress.	
Engineering	**Title: supervision** Implement the organizational appraisal system and ensure uniform application across the branch; provide the framework for objective development that is consistent for the division objectives; recognize high-performing employees and take corrective action for under/low-performing employees; establish/implement policy that promotes resource development. The incumbent will recognize high-performing employees by preparing and submitting at least two quarterly, annual or special award recognitions.	Resource management Leadership
Engineering	**Title: supervisory, first level supervisor with technical responsibilities** Select appropriate general use supervisory objective, balanced scorecard initiative, and contributing factors.	
Engineering	**Title: technical policy and direction, technical advisor** Implement division technical policy and direction for the branch and develop and implement additional functional area policy and direction to enhance branch effectiveness, as required; advise and lead personnel in the implementation of division and branch policies. • Develop a self-assessment tool for design engineering implementation and get concurrence from subject matter experts by January 31st. By March 31st, train 30% of staff and chief engineers on how to use of statistical testing tool. • Develop methodology to assess the implementation of technical policy in the division technical disciplines • Conduct a minimum of X policy implementation assessments.	Technical proficiency Communication Critical thinking

COMMUNICATING PROJECT OBJECTIVES (continued)

Functional Area	Objectives	Contributing Factors
Financial management	**Target: supervisor, financial management** Effectively balances workload with organizational needs and mission requirements to meet customer commitments and financial execution standards. Evaluates financial requirements of the program and projects assigned, and provides timely business advice and financial guidance to program management and functional team members. Plans work, sets short- and long-term objectives and prioritizes competing tasks to fit within schedule requirements. Effectively assigns work to staff and leverages all available automated tools to improve efficiency of financial performance and satisfy reporting requirements. Employee networks with functional staff and across organizational boundaries to leverage available knowledge and implement financial best practices.	Technical proficiency Critical thinking
Financial management	**Target: supervisor, financial management** Communicates and aligns strategies and financial policies within the organization. Identifies objectives that support organizational balanced scorecard map and establishes initiatives to demonstrate compliance. Sets specific performance goals, effectively communicates benefits to the financial team, and elicits participation and performance through a variety of means. Drives transformation by working across functional boundaries and cultivates effective working relationships with other team leaders within the department and on the functional staff.	Leadership Cooperation Teamwork
Financial management	**Target: supervisor, financial management** Employee provides timely, accurate, and concise program financial analysis to the organization. Effectively integrates analysis developed by other functional experts. Ensures all levels of management within and outside the organization are provided with accurate and comprehensive financial reporting. Promptly identifies funding challenges to upper level management. Reviews and ensures accuracy and completeness of all financial reporting.	Communication Customer focus
Functional staff office	**Target: functional experts** Leads, integrates, and implements new acquisition improvement initiatives and policy across the organization. Responsible for organizing, training, and equipping managers to perform acquisition functions to acquire and sustain project assets on time and on cost. Defines, develops, and matures processes/guidance	Technical proficiency Critical thinking

COMMUNICATING PROJECT OBJECTIVES (continued)

Functional Area	Objectives	Contributing Factors
	incorporating new acquisition initiatives. Develops improved tools, processes, and procedures to assist groups in the efficient execution on their assigned programs, and provides training for implementation. Establishes, coordinates, and implements policy/strategies/procedures for managing the workforce. Recruits, trains, and provides career development to all levels of the workforce.	
Functional staff office	**Target: functional staff office** Champion a compelling strategy that embodies acquisition and sustainment policies and communicates it to members of the organization. Devises and executes plans that align with organization's balance scorecard objectives and makes positive impacts to the organization. Sets goals and exercises delegation to accomplish them. Drives transformation by thinking and working across boundaries and cultivating solid relationships with other leaders.	Leadership Cooperation Teamwork
Functional staff office	**Target: supervisor, customer service** Provide timely, accurate, and concise program status reports in compliance with applicable policy and procedures. Effectively integrates analysis for internal and/or external customers. Clearly communicates to customers about functional initiatives and issues. Represents the organization on functional issues.	Communication Customer focus
Human resources	**Target: human resources** Develops and recommends an integrated, executable strategic workforce management program for the organization, addressing current and future requirements. Develops effective strategies to implement human resource programs, balancing departmental objectives within the context of division and higher level constraints and changing HR program direction/requirements. Institutes meaningful progress/success measurements to continually assess strategy effectiveness, adjusting as required.	Critical thinking Technical proficiency
Logistics	**Target: logistics leaders** Champions a compelling strategy that embodies acquisition and sustainment policies and communicates it to members of the organization. Devises and executes plans that align with organization's balance scorecard objectives and makes positive impacts to the organization. Sets goals and exercises delegation to accomplish them. Drives transformation by thinking and working across boundaries and cultivating solid relationships with other leaders.	Leadership Cooperation Teamwork

COMMUNICATING PROJECT OBJECTIVES (continued)

Functional Area	Objectives	Contributing Factors
Program management	**Target: program manager** Effectively plan and execute assigned projects to meet commitments to customers and organizational financial execution standards. Evaluate and employ advice and recommendations of functional representatives assigned to project teams. Apply current organizational tools/processes to streamline acquisition plans, manage risks proactively, design in supportability, and design in test and evaluation. Resolve issues on current programs in a timely manner. Manage and assess cost/schedule/technical performance and sustainment of assigned programs. Define and implement timely corrective actions when necessary to deliver as promised.	Technical proficiency Critical thinking
Financial management	**Target: project leaders** Champion a compelling strategy that embodies acquisition and sustainment policies and communicate it to members of the organization. Devise and execute plans that align with organization's balance scorecard objectives and make positive impacts to the organization. Set goals and exercise delegation to accomplish them. Drive transformation by thinking and working across boundaries and cultivating solid relationships with other leaders.	Leadership Cooperation Teamwork
Acquisitions office	**Target: first-line supervisor** Provide timely, accurate, and concise program status reports in compliance with applicable policy and procedures. Effectively integrate analysis provided by assigned functional representatives. Clearly communicates program successes and issues. The reports objectively reflect program performance as compared to program baselines or performance standards. The reports will be submitted by established deadlines and rarely need rework.	Communication Customer focus
For general use: Manager	**Target: general manager** Balances and manages workload within the unit considering available resources, difficulty of requirements and assignments, and the experience and ability of staff. Develops realistic financial plans for the unit and ensures effective use of resources to achieve unit goals and objectives. Identifies and manages unit priorities, recognizing and implementing necessary change. Builds and leads an effective team to produce quality products, meet organizational goals and customer needs.	Resource management

COMMUNICATING PROJECT OBJECTIVES (continued)

Functional Area	Objectives	Contributing Factors
For general use: Manager	**Target: management objective** Manage work effort to meet program objectives. Establishes operating guidelines, program direction and priorities, refining to meet changing situations. Balances efforts among competing projects and units, considering available resources, difficulty of projects, and the experience and ability of staff. Develops realistic financial plans for the unit. Builds and leads an effective team to produce quality products, and meet organizational goals and customer needs.	Resource management

From the sample in the preceding sections, it is obvious that a great deal of communication responsibilities lie with the project leadership. The list below re-emphasizes the leadership criteria presented earlier in Chapter 1:

- Technical proficiency
- Critical thinking
- Communication skills
- Cooperation ability
- Coordination skills
- Teamwork skills
- Resource management proficiency
- Leadership ability

GUIDELINES FOR TRIPLE C COMMUNICATION

Communication channels must be kept open throughout the project life cycle. In addition to internal communication, appropriate external sources should also be consulted. The project management must do the following:

- Exude optimism about the project and avoid open criticism of the project
- Exhibit positive attitude and constructive encouragement of personnel
- Utilize a communication responsibility matrix as a tool for effectiveness
- Facilitate multi-channel communication interfaces
- Identify and embrace internal and external communication avenues
- Avert or resolve organizational and project conflicts
- Encourage both formal and informal communication linkages

SUGGESTED CONTENTS OF TRIPLE C COMMUNICATION

The communication function of Triple C involves making all those concerned aware of project requirements and progress. To summarize earlier sections, those who will be affected by the project directly or indirectly, as direct participants, stakeholders, or beneficiaries, should be explicitly informed as appropriate regarding the following items:

- The scope of the project
- The personnel contribution required
- The expected cost of the project, in terms of both human efforts and materials
- The merits of the project
- The project implementation plan
- The potential adverse effects of the project, if it should fail
- The alternatives, if any, for achieving the project goal
- The potential direct and indirect benefits of the project, to the organization as well as individuals

With the above guidelines for Triple C communication, the risk of misinterpretation will be minimized within the project system. Clear communication leads to full understanding, which enhances the potential for cooperation. If we align our resources, we can improve communication, cooperation, and coordination, resulting in higher productivity, more effectiveness, better efficiencies, and manifestation of the organization's strategic plan. The next chapter discusses project cooperation and the processes for securing it.

4 Project Cooperation

The only thing that will redeem mankind is Cooperation.

Bertrand Russell, British philosopher, 1872–1970

The Chinese proverb presented at the beginning of Chapter 2 illustrates the importance of involving participants as a mechanism for achieving cooperation. Naturally, when we "understand," our platform of cooperation will increase. Involvement grounded in cooperation facilitates deep understanding and a genuine expression of cooperation. To cooperate is to organize into effective teams to accomplish work. This chapter presents organizational structures and cross-cultural teaming as the basis for project cooperation and collaboration. To achieve full and willing cooperation, there must be an assessment of personal trade-offs vis-à-vis organizational trade-offs. It should be understood that there is a difference between compliance and cooperation. The former implies command and control, while the latter implies companionship and partnership.

COMMON CAUSES OF LACK OF COOPERATION

Lack of cooperation is frequently cited as a reason for project failure. Cooperation is a human attribute. It cannot be achieved through technical modeling or implementation. It is important to understand and appreciate the common causes of lack of cooperation so that remedial measures can be developed. Some the causes of not achieving cooperation are as follows:

1. Lack of knowledge and understanding of project requirements
2. Incongruent context between the sides involved in the project
3. Lack of trust, perhaps due to previous experiences
4. Fear of the unknown
5. Lack of positive precedent
6. Indolence and lackadaisical attitude toward project
7. Absence of project management office (PMO) as a rallying point
8. Misconceptions about time, cost, and performance expectations
9. Nonconducive organizational structure
10. Lack of cultural integration

The dangers of skipping cooperation or taking it for granted can be seen in the increasing numbers of project failures cited in the news. Cryptic communication will result in no cooperation, which will result in disjointed coordination and ineffective use of resources. Lack of inherent cooperation is the main reason that project implementations sputter, without sound levels of efficiency and effectiveness. Greater cooperation enhances responsibility and accountability.

COOPERATION FOR PROJECT PLANNING

Planning is the beginning of progress! The key to a successful project is good planning. Project planning provides the basis for the initiation, implementation, and termination of a project. It sets guidelines for specific project objectives, project structure, tasks, milestones, personnel, cost, equipment, performance, and problem resolutions. An analysis of what is needed and what is available should be conducted in the planning phase of new projects. The availability of technical expertise within the organization and outside the organization should be reviewed. If subcontracting is needed, the nature of the contract should undergo a thorough analysis. The question of whether or not the project is needed at all should be addressed. The "make," "buy," "lease," "subcontract," or "do nothing" alternatives should be compared as a part of the project planning process.

When selecting a project and its associated work packages, planning should be done in an integrative and hierarchical manner following the levels of planning outlined here:

- Supra-level (organizational level conveying overall strategic direction)
- Macro-level (departmental level conveying operational direction)
- Micro-level (work station level conveying tactical direction for task accomplishment)

A plan must show vision, passion, and imagination. It must convey institutional vitality and vibrancy. A plan must contain a roadmap to success, bring results, be important to all, be relevant to organizational direction, and be sustainable. Moreover, a plan must contain performance metrics, include tracking processes, and contain assessment instruments. All these attributes help to secure cooperation for the plan.

Figure 4.1 shows the hierarchy of project planning for enhanced cooperation in a project environment. Supra-level planning deals with the big picture of how the project fits overall and long-range organizational goals. Questions faced at this level concern potential contributions of the project to the survival of the organization, effect on the depletion of company resources, required interfaces with other projects within and outside the organization, risk exposure, management support for the project, concurrent projects, company culture, market share, shareholder expectations, welfare of employees, and financial stability.

Macro-level planning deals with decisions that address the overall planning within the project boundary. The scope of the project and its operational interfaces should be addressed at this level. Questions faced at the macro level include goal definition, project scope, availability of qualified personnel, resource availability, project policies, communication interfaces, budget requirements, goal interactions, deadline, and conflict resolution strategies.

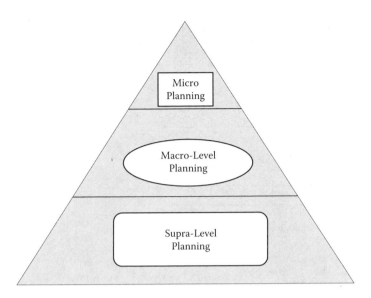

FIGURE 4.1   Hierarchy of project planning.

Micro-level planning deals with detailed operational plans at the task levels of the project. Definite and explicit tactics for accomplishing specific project objectives are developed at the micro level. The concept of management by objective (MBO) may be particularly effective at this level. MBO permits each project member to plan his or her own work at the micro level. Factors to be considered at the micro level of project decisions include scheduled time, training requirement, tools required, task procedures, reporting requirements, and quality requirements.

Whichever level is addressed, planning is an ongoing process that is conducted throughout the project life cycle. Initial planning may relate to overall organizational efforts. This is where specific projects to be undertaken are determined. Subsequent planning may relate to specific objectives of the selected project.

ORGANIZATIONAL STRUCTURING FOR COOPERATION

The organizational structure adopted for a project must produce an environment that is conducive for project cooperation. Both the traditional and contemporary structures should be explored for organizing complex projects. The traditional organizational structures such as matrix, product, and functional can be supplemented by unconventional adaptive structures such as those presented in Figures 4.2, 4.3, 4.4, 4.5, 4.6, and 4.7. Figure 4.2 shows a structure that captures the essence of time-dependent chronological relationships among activities (ACT), jobs, and tasks in a project.

Figure 4.3 shows sequential relationships among activities. The relationships do not necessarily have specific time dependency among them. Figure 4.4 presents what the author refers to as a "Bubble Wrap" structure. In the bubble wrap structure, teams and functions wrap around the overall project goal in a protective concept. In other words, everyone rallies around the project goal. This depends on the assumption that

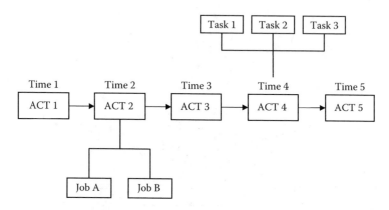

FIGURE 4.2   Chronological organization structure.

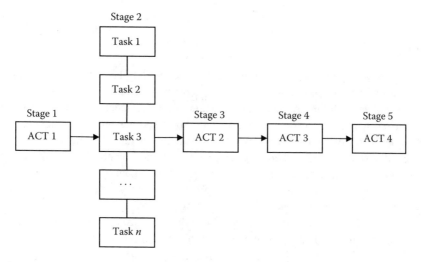

FIGURE 4.3   Sequential organization structure.

adequate communication had taken place to get everyone's "buy-in" into the project goal. Figure 4.5 conveys the notion that project power should center most appropriately within the project group, while leadership provides the foundational direction for the team's operation. By contrast, Figure 4.6 shows an autocratic structure where most of the power and decision making are centered with the leadership. Very few projects nowadays can succeed under this structure. Even military organizations, which had traditionally exercised "command and control" rigidly, are beginning to embrace the process of group decision making. This confirms the amorphous and dynamic state of contemporary projects. Figure 4.7 shows the interactive path for the transfer of functional elements from one project to another. Only specific relevant elements should be transferred from one project to another because "one size does not fit all" in project execution. Each project will have its own needs and nuances. The specific approach followed for each one should be tailored to the prevailing

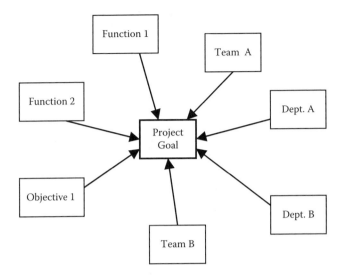

FIGURE 4.4 Bubble wrap organization structure.

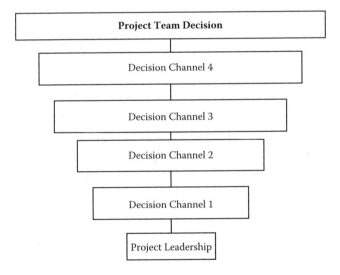

FIGURE 4.5 Political project structure.

circumstances and ambient environment of the project. This is particularly important since the customer base for a project shifts from one form to another.

MATRIX ORGANIZATION STRUCTURE

Matrix organization represents a mixture of pure project organization and functional organization. It permits both vertical and horizontal flows of information.
The matrix model is sometimes called a multiple-boss organization. It is a model
that is becoming increasingly popular as the need for information sharing increases.

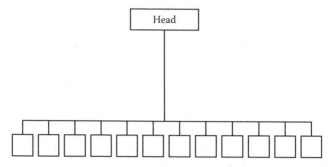

FIGURE 4.6 Autocratic organization structure.

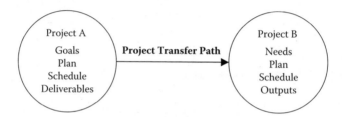

FIGURE 4.7 Internal and external organizational linkage.

Engineering projects, for example, require the integration of specialized skills from different functional areas. Under matrix organization, projects are permitted to share physical resources as well as managerial assets. An example of a project organized under the matrix model is shown in Figure 4.8. Figure 4.9 shows an alternate matrix view of the project. The project management office (PMO) may be responsible for all projects in the organization, but each project may have a home in a specific department within the organization; for example, the engineering department. Resources for the project are pooled from various departments across the organization.

A matrix organization is suitable where there is multiple managerial accountability and responsibility for a job function. There are usually two chains of command, horizontal and vertical, dealing with functional and project line. The project line in the matrix is usually of a temporary nature, while the functional line is more permanent. The matrix organization is quite dynamic, with its actual structure being determined by the prevailing project scenarios. The matrix organization has several advantages and some disadvantages. The advantages are as follows:

1. **Consolidation of objectives:** The objectives of the task at hand are jointly shared and pursued by multiple departments.
2. **Efficient utilization of resources:** The allocation of company resources is more streamlined. Manpower and equipment can be allocated at the most suitable usage level jointly among departments working toward a common goal.
3. **Free flow of information:** Since departments are cooperating rather than competing, there is an unhindered flow of common information both vertically and horizontally in the matrix structure.

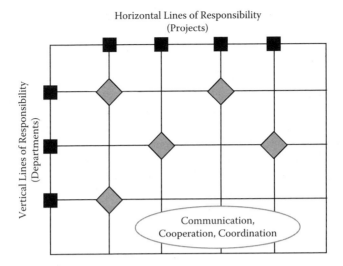

FIGURE 4.8   Matrix organization structure.

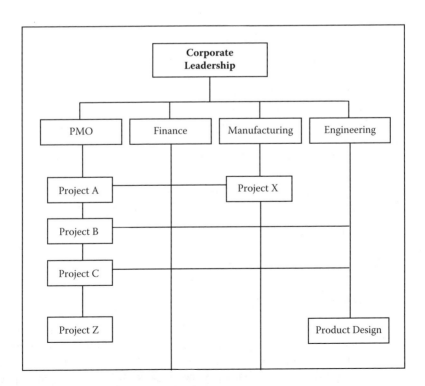

FIGURE 4.9   Alternate matrix organization structure.

4. **Interpersonal contacts:** The joint responsibility for projects creates an atmosphere of functional compatibility. Good working relationships that develop under one matrix structure become useful in other projects.
5. **High morale:** The success achieved on one project effort motivates workers to cooperate on other projects.
6. **Lateral functional interactions:** The multiple responsibilities for projects allow workers to be exposed to other functional activities and, thereby, permit smooth transition to other departments should that become necessary.
7. **Postproject interactions:** The matrix structure allows continuity of functions after project conclusion. The functional departments simply redirect their efforts to other responsibilities. Unlike a project organization structure where project shutdown could necessitate layoffs, the matrix structure makes a provision for returning to regular responsibilities.

The disadvantages of the matrix organization are as follows:

1. **Multiple bosses:** A major disadvantage of the matrix structure is the fact that workers report to two bosses on a given project. Playing one boss against another is a potential problem in a matrix structure.
2. **Power struggle:** A power struggle between the bosses may adversely affect the coordination of a project.
3. **Complexity of structure:** The number of managers and personnel involved in a given project can easily be confusing. Difficulties can arise with respect to monitoring and controlling personnel activities. Other potential problems are obstruction of information, slow response time, difficulty in resolving conflicts, unclear channels for supervision, and incompatibility of policies and procedures.
4. **Overhead cost:** By doubling up the chain of command, the matrix structure leads to higher interdepartmental and intradepartmental overhead costs. However, as productivity gains are realized, the overhead costs may become negligible.
5. **Conflicting priorities:** Since multiple responsibilities is a major characteristic of a matrix organization, it is sometimes difficult to determine which responsibility has higher priority. Each functional manager may view his own direct responsibilities as having higher priority than other project responsibilities.

Despite its disadvantages, the matrix organization is widely used in practice. Its numerous advantages tend to outweigh the disadvantages. Besides, the disadvantages can be overcome with proper managerial effort.

ORGANIZATIONAL BUREAUCRACY

Bureaucratic organizations are usually too inflexible and unresponsive for the needs of contemporary projects. Projects in the 21st century are diverse, multifaceted, and environmentally sensitive. It is important to abandon the age-old reliance on bureaucracy to get things done. Contemporary teams must be empowered to pursue self-organization and self-governance to accomplish work based on their "bird's eye view" of the factors that directly influence a project. The traditional bureaucratic processes are often too far removed from the realities of a project. The Triple C approach encourages creating a flatter organizational structure where ideas and functional contributions flow within a project team. If a team is built under the principles of communication and friendly cooperation, roles and responsibilities will evolve to accomplish project goals in an efficient and expeditious manner.

PROJECT CHARTER AND COOPERATION

A project charter represents a document of cooperation among team members on a project. A well-designed charter will facilitate agreement and cooperation throughout the rank and file of a project organization. A project charter is one of the first steps in executing a project. It more explicitly communicates the project identification, definition, and initiation steps in a project hierarchy. The charter can make or break a successful project. When fully understood, a charter can help solidify team focus, effectiveness, and motivation.

OVERVIEW OF A CHARTER

The purpose of a charter is to define at a high level what the project is about, what the project will deliver, what resources are needed, what resources are available, and how the project is justified. The charter also represents an organizational commitment to dedicate the time and resources to the project. The charter should be shared with all stakeholders as a part of the communication requirement of the Triple C approach. Cooperating stakeholders will not only signoff on the project, but also make personal pledges to support the project.

It is desired for a charter to be brief. Depending on the size and complexity of a project, the charter should not be more than two to three pages. Where additional details are warranted, the expatiating details can be provided as addenda to the basic charter document. The longer the basic charter, the less the likelihood that everyone will read and understand the contents. So, brevity and conciseness are desired virtues of good project charters. The charter should succinctly establish the purpose of the project, the participants, and general vision for the project.

The project charter is used as the basis for developing project plans. While it is developed at the outset of a project, a charter should always be fluid. It should be reviewed and updated throughout the life of the project. The components of the project charter are summarized here:

- Project overview
- Project goals
- Impact statement
- Constraints (time, cost, performance)
- Assumptions
- Project scope
- Project objectives
- Financial implications

- Project approach (policies, procedures)
- Project organization

The project charter does not include the project plan. Planning documents, which may include project schedule, quality plan, staff plan, communication hierarchy, financial plan, and risk plan, should be prepared and disseminated separately from the charter.

- Project overview
  - The project overview provides a brief summary of the entire project charter. It may provide a brief history of the events that led to the project, an explanation of why the project was initiated, a description of project intent and the identity of the original project owner.
- Project goals
  - Project goals identify the most significant reasons for performing a project. Goals should describe improvements the project is expected to accomplish along with who will benefit from these improvements. This section should explain what various benefactors will be able to accomplish as a result of the project. Note that the Triple C approach requires these details to secure cooperation.
- Impact statement
  - The impact statement identifies the influence the project may have on the business, operations, schedule, other projects, current technology, and existing applications. While these topics are beyond the domain of this project, each of these items should be raised for possible action.
- Constraints and assumptions
  - Constraints and assumptions identify any deliberate or implied limitations or restrictions placed on the project along with any current or future environment the project must accommodate. These factors will influence many project decisions and strategies. The potential impact of each constraint or assumption should be identified.
- Project scope
  - Project scope defines the operational boundaries for the project. Specific scope components are the areas or functions to be impacted by the project and the work that will be performed. The project scope should identify both what is within the scope of the project and what is outside the scope of the project.
- Project objectives
  - Project objectives identify expected deliverables from the project and the criteria that must be satisfied before the project is considered complete.
- Financial summary
  - The financial summary provides a recap of expected costs and benefits due to the project. These factors should be more fully defined in the cost-benefit analysis of the project. Project financials must be reforecast during the life of the effort.

- Project approach
  - Project approach identifies the general strategy for completing the project and explains any methods or processes, particularly policies and procedures that will be used during the project.
- Project organization
  - The project organization identifies the roles and responsibilities needed to create a meaningful and responsive structure that enables the project to be successful. Project organization must identify the people who will play each assigned role. At minimum, this section should identify who plays the roles of project owner, project manager, and core project team.
  - A project owner is required for each project.

    This role must be filled by one or more individuals who are the fiscal trustee(s) for the project to the larger organization. This person considers the global impact of the project and deems it worthy of the required expenditure of money and time. The project owner communicates the vision for the effort and certifies the initial project charter and project plan. Should changes be required, the project owner confirms these changes and any influence on the project charter and project plan. When project decisions cannot be made at the team level, the project owner must resolve these issues. The project owner must play an active role throughout the project, especially ensuring that needed resources have been committed to the project and remain available.
  - A project manager is required for each project.

    The project manager is responsible for initiating, planning, executing, and controlling the total project effort. Members of the project team report to the project manager for project assignments and are accountable to the project manager for the completion of their assigned work.

CULTURAL AFFAIRS IN MULTINATIONAL PROJECTS

Cultural faux pas can ruin or hinder project cooperation. For example, in African cultures, it is taboo to go "Dutch" with guests in a restaurant. You are either capable of picking up the tab or not at all. Splitting a check at the end of a business dinner leaves a sour taste in the psyche of an African. Yet, many Western organizations and governments forbid or at least discourage meal favors. But many other parts of the world take delight in and highly subscribe to hosting meals for interorganizational associates. Figure 4.10 illustrates an example of how global project cooperation might be enhanced through Triple C. Communication, cooperation, and coordination serve as the three legs holding up a joint global effort. Nowadays, the world has conceptually shrunk so much that speedy communication must be effected across wide geographical locations. The world continents of North America, South America, Africa, Europe, Asia, Australia/Oceania, and Antarctica all have their zonal challenges for global project communication. Consequently, special considerations and efforts must be directed at achieving effective communication.

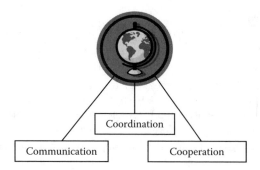

FIGURE 4.10   Global project cooperation.

WORKFORCE CULTURAL ATTRIBUTES FOR COOPERATION

Global cooperation has become more important in recent years because of the decreasing gaps in country-to-country communication and interactions. Global outsourcing of projects is a good case in point. Most outsourcing points are located in developing and underdeveloped nations. These locations often have authoritarian cultures that contain norms that the Western world would find unacceptable. A cultural bridge usually is missing between the developed nations and the developing nations with respect to outsourced projects. Thus, there are increasing cultural and economic disparities between global business partners in the selection of outsourcing points. Some of the local issues to be factored into outsourcing projects are summarized here:

- Poverty
- Pollution
- Disease
- Poor hygiene
- Political oppression
- Gender biases
- Economic and financial scams
- Wealth inequities
- Sexual subjugation of women
- Social and sexual permissiveness among elites

Project representatives posted to culturally disparate locations around the world are shocked by the cultural differences that they witness. In some cases, they maintain a "laissez faire" and hands-off attitude. But there have also been cases where some of these representatives take advantage of the loose culturally acceptable social contacts that could be to the detriment of a multinational project.

   In a culturally sensitive workforce, the specific levels and structure of the needs may be drastically different from the typical mode. In a non-Western outsource point (sink), most workers will be at the basic level of physiological needs; and there will exist cultural constraints on moving from one level to the next higher level. This fact has an implication on how cultural interfaces can occur between hosts and guests in multinational project outsourcing situations. A culture-induced disparity in hierarchy of needs implies that an outsourcing company should recognize how to properly deal with the typical needs of the workforce at each level. The typical hierarchy of needs of project team members is outlined here (from Maslow, 1943).

1. **Physiological Needs:** The needs for the basic necessities of life, such as food, water, housing, and clothing (survival needs). This is the level where access to wages is most critical. The basic needs are of primary concern in an outsource location.
2. **Safety Needs:** The needs for security, stability, and freedom from threat of physical harm.
3. **Social Needs:** The needs for social approval, friends, love, affection, and association. Industrial outsourcing may bring about better economic outlook that may enable each individual to be in a better position to meet his or her social needs.
4. **Esteem Needs:** The needs for accomplishment, respect, recognition, attention, and appreciation. These needs are important not only at the individual level, but also at the organizational level.
5. **Self-Actualization Needs:** These are the needs for self-fulfillment and self-improvement. They also involve the stage of opportunity to grow professionally. Project outsourcing may create opportunities for individuals to assert themselves socially and economically.

A classic anecdotal example of hierarchy-induced lack of cooperation was witnessed by the author once in a Third World country. At a local hotel, a Western businessman went to the front desk to complain that the air-conditioner in his room was not working and requested that the hotel personnel do something about it urgently. The front desk clerk called a maintenance man over and instructed him to take care of the service call. The maintenance man whispered to the clerk in disgust, in his local language, that "what is this man complaining about? Just look at him. He is asking for air-conditioning. In my own home, I have not even seen electricity in two weeks, not to talk of air-conditioning." Needless to say, the service call was not attended to that evening. In this case, communication took place quite all right. But the context under which cooperation was solicited was completely misplaced. And, of course, coordination of efforts failed because there was an absence of cooperation.

TECHNOLOGY FACTORS FOR PROJECT OUTSOURCING

In general, information technology (IT) for project outsourcing purposes has definite requirements that may be viewed as qualitative or quantitative and may be divided into technology factors and people factors as summarized here:

TECHNOLOGY FACTORS

- Commitment of investment in IT programs
- Hardware installation
- Hardware and software maintenance

- Frequency of enhancements and upgrades
- Reciprocity with other information outlets

PEOPLE FACTORS

- Recruitment of skilled workforce
- Training of the workforce for specific outsourcing needs
- Retention programs for experienced workforce
- Adaptability of the workforce to new modes of business operations

The level of performance with respect to these factors will affect the feasibility and probability of success of outsourcing implementation. The feasibility of an outsourcing proposal may be determined from several perspectives covering cultural, social, administrative, technical, and economical issues. Typically, the technical and economic aspects get more attention. But in reality, it may be the cultural and social aspects that will determine the success of industrial outsourcing.

CULTURAL AND SOCIAL COMPATIBILITY

Cultural infeasibility is one of the major impediments to project outsourcing in an emerging economy. The business climate is very volatile. This volatility, coupled with cultural limitations, creates problematic operational elements in a developing country. The pervasiveness of Internet information overwhelms the strict cultural norms in most developing countries. The cultural feasibility of information-based outsourcing needs to be evaluated from the standpoint of where information originates, where it is intended to go, and who comes into contact with the information. For example, the revelation of personal information is viewed as taboo in many developing countries. Consequently, this impedes cooperation for the collection, storage, and distribution of workforce information that may be vital to the success of multinational projects.

For project outsourcing to be successfully implemented in such settings, assurances must be incorporated into the hardware and software implementations so as to soothe the concerns of the workforce. Accidental or deliberate mismanagement of information is a more worrisome aspect of IT than it is in the Western world, where enhanced techniques are available to correct information errors. What is socially acceptable in the outsourcing culture may not be acceptable in the receiving culture; and vice versa.

ADMINISTRATIVE COMPATIBILITY

Administrative or managerial feasibility involves the ability to create and sustain an infrastructure to support an operational goal. Should such an infrastructure not be in existence or unstable, then we have a case of administrative infeasibility, which subsequently can impede cooperation. In developing countries, a lack of trained manpower precludes a stable infrastructure for some types of project outsourcing. Even where trained individuals are available, the lack of coordination makes it almost impossible to achieve a collective and dependable workforce. Projects that are designed abroad for implementation in a different setting frequently get bogged down when imported into a developing environment that is not compatible for such project implementations.

Differences in the perception of ethics are also an issue of concern in an outsource location. A lack of administrative vision and limited managerial capabilities limit the ability of outsource managers in developing countries. Both the physical and conceptual limitations on technical staff lead to administrative infeasibility that must be reckoned with. Overzealous entrepreneurs are apt to jump on opportunities to outsource production without a proper assessment of the capabilities of the receiving organization. More often than not outsourcing organizations don't fully understand the local limitations. Some organizations take the risk of learning as they go along, without sufficient advance preparation.

TECHNICAL COMPATIBILITY

Hardware maintenance and software upgrade are, perhaps, the two most noticeable aspects of technical infeasibility of information technology in multinational projects in a developing country. A common misconception is that once your IT is installed and all its initial components are in place, the system is available for its entire life cycle. This is very far from the truth. The lack of proximity to the source of hardware and software enhancement makes this situation particularly worrisome in multinational projects in underdeveloped countries. The technical capability of the personnel as well as the technical status of the hardware must be assessed in view of the local needs. Doing an overkill on the infusion of IT just for the sake of keeping up is as detrimental as doing nothing at all; and it can impede long-term project cooperation.

IT ROLE IN OVERSEAS PROJECT COOPERATION

The use of modern information communication technology (i.e., the Internet) has produced remarkable changes in the ways we communicate and conduct business. We are witnessing the emergence of virtual cooperating teams, which consist of geographically dispersed coworkers whose interaction is facilitated through information communication technologies. This is made possible by new and high levels of global connectivity. Significant advances have been seen within the past few years in how people use the latest information technology to conduct both official and personal business. In economically developed countries, most enterprises intricately entwine telecommunications in their day-to-day operations and management. With such integration, the Internet constitutes both a threat and a promise. The threats are due to the ways in which electronic communication has significantly altered the way information is captured, stored, manipulated, disseminated, and used. As a result, organizations are vulnerable to potential misuses and abuses—either through accidental incidents or deliberate intents. The promise offered by the Internet relates to the potential to achieve notable enhancements in business processes. The magnitudes of the promises and threats that the Internet presents are particularly pronounced in developing countries, where there is less infrastructure to support it and fewer policies to control its use in business and personal operations.

The Internet has become a global tool for project cooperation. It is seen not only as an instrument of information handling, but also as a medium for technological advancement. The latter is important for developing nations because of the increasing reliance on importation of outsourced multinational projects; a key result of globalization. If the Internet can be the vehicle for enhanced international trade, then it will assist countries to become more export oriented and be more competitive in international projects. With the aid of the Internet, more and more developing countries are exporting "services" rather than physical products. Manufacturers in industrialized countries can quickly conduct an assessment of potential labor forces available in developing countries to determine outsource targets. Modern information technology has facilitated the following types of multinational project transactions and cooperation:

- Transactions for a service, which is completed entirely on the Internet from selection to purchase and delivery.
- Transactions involving "distribution services" in which products, whether goods or services, are selected and purchased online but delivered by conventional means.
- Transactions involving telecommunication or routing function of business information.

In order to achieve a successful marriage of IT transactions and existing business functions, a company must understand the respective characteristics of both. Online transactions are characterized by specific functionalities such as online real-time product presentation (electronic catalog), order entry (electronic shopping), product distribution, customer service, product support, and data acquisition. These functionalities create a business environment that is:

- Fast paced
- Expensive to initiate
- Requires formal training programs

Not all of the above are possible (or feasible) within the existing electronic infrastructure of developing countries. Consequently, there are possible technical and cultural pitfalls such as:

- Errors in communication translation
- Errors in communication transmission
- Errors in communication assimilation

WORKFORCE INTEGRATION STRATEGIES

Any overseas project outsourcing requires an adaptation from one form of culture to another. The implementation of a new technology to replace an existing (or a nonexistent) technology can be approached through one of several cultural adaptation options. The following are some suggestions:

- **Parallel Cooperation:** The host culture and the guest culture operate concurrently (side by side); with mutual respect on either side.
- **Adaptation Cooperation:** This is the case where either the host culture or the guest culture makes a conscious effort to adapt to each others' ways. The adaptation often leads to new (but not necessarily enhanced) ways of thinking and acting.
- **Superimposition Cooperation:** The host culture is replaced (annihilated or relegated) by the guest culture. This implies cultural imposition on local practices and customs. Cultural incompatibility, for the purpose of business goals, is one reason to adopt this type of interface.
- **Phased Cooperation:** Modules of the guest culture are gradually introduced to the host culture over a period of time.
- **Segregated Cooperation:** The host and guest cultures are separated both conceptually and geographically. This used to work well in colonial days. But it has become more difficult with modern flexibility of movement and communication facilities.
- **Pilot Cooperation:** The guest culture is fully implemented on a pilot basis in a selected cultural setting in the host country. If the pilot implementation works with good results, it is then used to leverage further introduction to other localities.

CULTURAL HYBRIDS FOR PROJECT COOPERATION

The increased interface of cultures through project outsourcing is gradually leading to the emergence of hybrid cultures in many multinational projects. A hybrid culture derives its influences from diverse factors, where there are differences in how the local population views education, professional loyalty, social alliances, leisure pursuits, and information management. A hybrid culture is, consequently, not fully embraced by either side of the cultural divide. This creates a big challenge to managing outsourced projects.

WORKFORCE EDUCATION AND OUTSOURCING

Education should play a significant mitigation role in addressing cultural challenges facing multinational outsourced projects. Formal education introduces the workforce to the cultures of the world. Informal education (e.g., a refresher course) provides operating guidelines for professionals traveling overseas to coordinate multinational projects. Suggested training and education strategies for multinational projects include:

- Advance briefings
- International exchange programs for students, personnel, managers, and administrators
- Formal lectures and courses
- Training seminars

- Bilateral cultural meetings
- Round-table discussions involving government, business, industry, and academia
- Hands-on exercises
- Role playing activities
- Open debates designed to iron out and explain unique cultural traits
- Apprenticeship trips or excursions

TECHNOLOGY TRANSFER IN MULTINATIONAL PROJECTS

Technology transfer for multinational project purposes can be achieved in various forms. Three technology transfer modes are recommended here to illustrate basic strategies for getting one outsourced product from one point (technology source) to another point (technology sink). Technology can be transferred in one or a combination of the following strategies:

1. Transfer of complete project products
2. Transfer of project procedures and guidelines
3. Transfer of project concepts, theories, and ideas

TRANSFER OF COMPLETE TECHNOLOGY PRODUCTS

In this case, a fully developed product is transferred from a source to a target. Very little product development effort is carried out at the receiving point. However, information about the operations of the product is fed back to the source so that necessary product enhancements can be pursued. So, the technology recipient generates product information, which facilitates further improvement at the technology source. This is the easiest mode of technology transfer and the most tempting. Developing nations are particularly prone to this type of transfer. Care must be exercised to ensure that this type of technology transfer does not degenerate into mere "machine transfer" without local intellectual contribution.

TRANSFER OF TECHNOLOGY PROCEDURES AND GUIDELINES

In this technology transfer mode, procedures (e.g., blueprints) and guidelines are transferred from a source to a cooperating target. The technology blueprints are implemented locally to generate the desired services and products. The use of local raw materials and manpower is encouraged for the local production. Under this mode, the implementation of the transferred technology procedures can generate new operating procedures that can be fed back to enhance the original technology, in a sort of reciprocal cooperation. With this symbiotic arrangement, a loop system is created whereby both the transferring and the receiving organizations derive useful benefits.

TRANSFER OF TECHNOLOGY CONCEPTS, THEORIES, AND IDEAS

This strategy involves the transfer of the basic concepts, theories, and ideas supporting a multinational project. The transferred elements can be enhanced, modified, or customized within local constraints to generate new technology or products. The local modifications and enhancements have the potential to generate an identical technology, a new related technology, or a new set of technology concepts, theories, and ideas. These derived products may then be transferred back to the original technology source under a reverse cooperating agreement. Transferred technology must be implemented to work within local limitations. Local innovation, patriotism, dedication, and cultural flexibility to adapt are required to make multinational technology transfer successful.

GUIDELINES FOR MULTINATIONAL TECHNOLOGY COOPERATION

This section presents three technology transfer guidelines for multinational technology cooperation.

- A multinational project should assess which technology products are most suitable for transfer to a given outsource location, based on the specific attributes of the project and specific factors at the receiving location.
- A multinational project should assess which technology procedures and guidelines can be expected to generate the greatest benefits when transferred from a given source to a given target, based on cultural, social, and technical factors at both organizations.
- A multinational project should assess which technology concepts, theories, and ideas have the highest potential for practical implementation.

In order to reach the overall goal of a successful technology transfer, it is essential that the most suitable technology be identified, transferred under the most conducive terms, implemented at the receiving organization in the most appropriate manner at the right time, and managed with full commitment to cooperative reciprocity.

The explicit pursuit of project cooperation implies a need to overcome the common causes of lack of cooperation as mentioned at the beginning of this chapter. But there are also unknown causes of lack of cooperation that may be intractable. While proactively attempting to overcome the assignable causes of lack of cooperation, one must also try not to engage in unconstructive acts that outwardly lead to a lack of cooperation. Where cooperation does not exist, a project manager can use Triple C communication approach to explicitly seek cooperation. Where cooperation already exists, the project team must still work hard to ensure that negating incidents are not permitted to develop. Where a neutral platform is the prevailing scenario, the project leadership must ensure that an imbalance does not throw the project into the noncooperating realm.

The next chapter presents coordination as the final stage of the Triple C model. Having concluded the communication and cooperation stages, coordination ensures that activities of participating teams are correlated across functional lines.

REFERENCE

Maslow, Abraham H., A theory of human motivation. *Psychological Review*, Vol 1, pp. 370–396, 1943.

5 Project Coordination

Let our advance worrying become advance thinking and planning.

Winston Churchill

Coordination is the key that really gets things done in a project; after the initial period of project trepidation (worrying and anxiety) has been transcended through "thinking and planning." No project system operates in isolation. Each project must interact within and outside of its scope of operations. There will be interactions with multiple subsystems across different organizational platforms. This calls for coordination at various stages of a project.

As an anecdotal example, highway safety advocates often cite a lack of coordination among state motor vehicle bureaus as a big reason high-risk traffic offenders fall through the cracks and commit interstate infractions again and again. There is often good communication and cooperation among the bureaus. But without the coordination phase to actualize the cooperative agreement, nothing gets done effectively. If any public safety undertaking is approached from a project management perspective, under the structure of Triple C, more effective operations can be achieved.

Project effectiveness is no longer measured by mere volumes of resources allocated to it. Success, in a contemporary project, is predicated on effective coordination of human resources, work processes, and tools. So, what is coordination? Coordination refers to working with other people. For example, collecting information requires coordinating with other people who own, protect, or manage the information. Frequently, we talk of synergy of a work team. Synergy does require a high degree of coordination. Project coordination can be defined as follows:

> Project coordination is a balanced choreography of teamwork across the various elements of a project organization and among several members of the project team.

Each team member must exhibit conscientious commitment to the project so that project harmony can be assured. If project communication is done properly as the first stage of Triple C, commitment of team members becomes more realizable.

INITIATING COORDINATION

Project coordination requires mutual understanding of the goal and schedule of the project. That mutual understanding emanates from proper Triple C communication and cooperation. A primary requirement for a project manager is to constructively influence workers during the coordination phase of a project. Coordination doesn't just happen. It must be effected through direct actions under the leadership of a capable project manager. Proclamations and memos are effective in initiating coordination points in a project. An example memo is presented below. This memo can serve multiple purposes as a tool for communication, cooperation, and coordination.

Date: 07/07/07

To: All Concerned (ALCON)

From: PMO

Subject: Coordination Memo to General Staff

I have just received *communication* from the vice president's office that our division will conduct a search for a new director of project management at our Sunny Oasis location. Functional areas of responsibilities for the new hire will include project planning, resource allocation, performance assessment, and training. This is a good sign for a stable future of our division. It also points out our company's commitment to achieve a sustainable increase in our market share, which will positively impact our staff retention and opportunities for new employment. I urge everyone to participate in supporting the recruitment effort once the national search begins.

Thanks for your continuing *cooperation* regarding organizational improvement initiatives. With this, we can have effective *coordination* of our respective efforts.

Figure 5.1 illustrates the call-to-order aspect of project coordination, after the phases of communication and cooperation have been accomplished. Coordination must be an iterative process. Each stage of coordination initiates the next coordination stage. For coordination to be sustainable, "bureaucratic acrobatics" and "administrative gymnastics" must be avoided. Bureaucratic acrobatics refers to unnecessary bureaucracy that encumbers the work process through pretentious activities that feign high skills and performance excellence. For example, pontification and long-winded styles of operation are symptoms of a process grounded in bureaucracy. Project bureaucrats tend to act in self-important ways, especially when not qualified to handle a process efficiently. Verbal acrobatics is the process of saying a lot without really saying much, thereby impeding coordination toward project progress. Similarly, administrative gymnastics refers to the process of dancing around the core

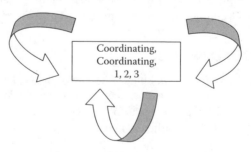

Coordinating,
Coordinating,
1, 2, 3

FIGURE 5.1   Initiation of coordination.

issues in a project. Focusing on tangential issues, rather than the "meat" of a project issue, impedes project coordination. Perambulatory maneuvering of tasks defeats the purpose of project coordination. Project team members must be watchful and wary of verbal acrobatics and verbal gymnastics and not be swayed by a project team member's personality.

Cohesive organization of efforts is required to achieve project coordination. Important elements of coordination include:

- Balancing of tasks
- Validation of time estimates
- Authentication of lines of responsibility
- Identification of knowledge transfer points
- Standardization of work packages
- Integration of project phases
- Minimization of change orders
- Mitigation of adverse impacts of interruptions
- Avoidance of work duplication
- Identification of team interfaces
- Verification of work rates
- Validation of requirements
- Identification and implementation of process improvement opportunities

ADAPTIVE PROJECT COORDINATION

Selecting appropriate and adaptive organizational structures for project coordination is essential for project success. Coordination must be effected across both the managerial and technical processes of a project as illustrated in Figure 5.2. In the figure, project coordination is modeled as a function of managerial process, technical process, adaptive project planning, and adaptive organizing. Coordination is the common thread that links all the requirements. A managerial process should be continually reviewed, modified, and improved to adapt to current business needs. The concept of continuous process improvement (CPI) facilitates an adaptive managerial process. Adaptive project planning involves using contingency planning to

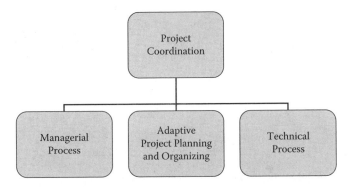

FIGURE 5.2 Coordination across managerial and technical processes.

respond to prevailing developments in the project environment. No plan should be cast in concrete. A plan should be developed to have avenues for modifications as new realities of the project are encountered. Adaptive organizing refers to the ability of the existing project structure to assume new physical forms and operational constructs based on contemporary needs of the organization. The technical process of an organization is the core asset that transforms concept into product. As new technologies develop, it is through an adaptive technical process that an organization can take advantage of the technologies. All of these require careful coordination, in deed, rather than just rhetoric.

COORDINATION FOR PROJECT SCHEDULING

Project scheduling is the most visible part of project management because a schedule indicates the beginning and end points of a project. A schedule is effected under the triple constraints of cost, time, and performance expectations. The Triple C approach presented in this book links triple constraints issues to Triple C processes to ensure project results. Figure 5.3 shows the top-down and bottom-up information flows between the triple constraints side and the Triple C side in a project system. Cost, schedule, and performance expectations must be linked to the communication, cooperation, and coordination requirements. Project cost is affected by team members' reactions to the communication, cooperation, and coordination aspects of a project. In the same way, performance is influenced by levels of communication, cooperation, and coordination. Finally, all these factors collectively determine how well a project schedule is formulated, executed, and sustained.

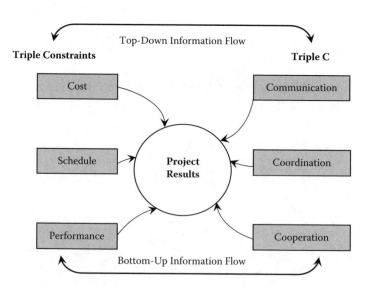

FIGURE 5.3   Coordination linkages to triple constraints.

COORDINATION FOR PROJECT CONTROL

Although project control is often listed as the last of the four major steps of project management (planning, organizing, scheduling, and control), it is actually a process that permeates all stages of a project. The coordination component of Triple C shows that communication and cooperation are essential for achieving appropriate control throughout a project. Figure 5.4 illustrates the link between project control and Triple C elements. Communication is required for every project team member to understand the essential contents of control. Cooperation is needed to obtain everyone's participation in providing inputs and data needed for control actions.

The three factors (time, budget, and performance) that form the basis for the operating characteristics of a project also help to determine the basis for project control. Project control is the process of reducing the deviation between actual performance and planned performance. To be able to control, we must be able to measure performance. Measurements are taken on each of the three components of project constraints: Time, performance, and cost. Figure 5.5 shows a flowchart of task data for project tracking and assessment of the data for feedback into project system for cost, time, and performance planning and feed-forward into resource coordination. The loop is closed by using the incoming measurement and assessment information to enhance the next phase of data collection.

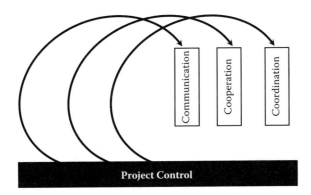

FIGURE 5.4   Application of Triple C for control.

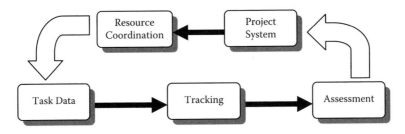

FIGURE 5.5   Coordination feedback control model.

Some of the factors that can cause a project to require control are as follows:

1. Time
   - Supply delays
   - Missed milestones
   - Delay of key tasks
   - Change of due dates
   - Unreliable time estimates
   - Increased need for expediting
   - Time-consuming technical problems
   - Impractical precedence relationships
   - New industry regulations that need time to implement
2. Performance
   - Poor quality
   - Poor mobility
   - Low reliability
   - Restricted access
   - Fragile components
   - Poor functionality of tools and equipment
   - Maintenance problems
   - Technical difficulties
   - Change order from client
   - High-risk implementation
   - Conflict of functional objectives
   - Resources not available when needed
   - Adjustments needed due to new technology
3. Cost
   - Cost overruns
   - Price changes
   - Incorrect bids
   - High labor cost
   - Budget revisions
   - High overhead rates
   - Poor cost reporting
   - Inadequate resources
   - Increase in scope of work
   - Increased delay penalties
   - Insufficient project cash flows

Project control may be handled in a hierarchical manner starting with the global view of a project and ending with the elementary level of unit performance. The hierarchy may be constructed as presented here:

Project system → Product line → Process → Product unit

FIGURE 5.6 Project development team coordination.

A product is project dependent. A process is product dependent. The performance of a product unit depends on the process from which the unit is made. Such a control hierarchy makes the control process more effective. Figure 5.6 shows the progression of coordination from project system through project, team, task, and product goal. Each coordination step requires communication and cooperation among all project elements.

CONTROL STEPS

Parkinson's Law states that a schedule will expand to fill available time and cost will increase to consume the available budget. Project control prevents a schedule from expanding beyond reason. Project control also ensures that a project can be completed within budget. A coordination-based project control process is presented here:

Step 1: Determine the criterion for control. This means that the specific aspect to measure should be determined.

Step 2: Set performance standards. Standards may be based on industry practice, prevailing project agreements, work analysis, forecasting, and so on.

Step 3: Measure actual performance. The measurement scale should be predetermined. The measurement approach must be calibrated and verified for accuracy. Quantitative and nonquantitative factors may require different measurement approaches.

Step 4: Compare actual performance with the specified performance standard. The comparison should be done objectively and consistently based on the specified control criteria.

Step 5: Identify unacceptable variance from expectation.

Step 6: Determine the expected impact of the variance on overall project performance.

Step 7: Investigate the source of the poor performance.

Step 8: Determine the appropriate control actions needed to overcome (nullify) the variance observed.

Step 9: Implement the control actions with total dedication.

Step 10: Ensure that the poor performance does not recur elsewhere in the project.

INFORMAL CONTROL PROCESS

Informal control refers to the process of using unscheduled or unplanned approaches to assess project performance and using informal control actions. Informal control requires unscheduled visits and impromptu queries to track progress. The advantages of thr informal control process are presented below:

ADVANTAGES OF INFORMAL CONTROL

- It allows the project manager to learn more about project progress.
- It creates a surprise element, which keeps workers on their toes.
- It precludes the temptation for "doctored" progress reports.
- It allows peers and subordinates to assume control roles.
- It facilitates prompt appraisal of latest results.
- It gives the project manager more visibility.

FORMAL CONTROL PROCESS

A formal control process deals with the process of achieving project control through formal and scheduled reports, consultations, or meetings. Formal control is typically more time consuming.

DISADVANTAGES OF FORMAL CONTROL PROCESS

- It can be used by only a limited (designated) group of people.
- It reduces the direct visibility of the project manager.
- It can impede the implementation of control actions.
- It encourages bureaucracy and "paper pushing."
- It requires a rigid structure.
- It is more time consuming.

Despite its disadvantages, formal control can be effective for industrial development projects because of the size and complexity of such projects. With a formal control process, project responsibilities and accountability can be pursued in a structured manner. For example, standard audit questions may be posed in order to determine current status of a project and establish the strategy for future performance. Examples of such questions are the following:

- Where are we today?
- Where were we expected to be today?
- What are the prevailing problems?
- What problems are expected in the future?
- Where will we be at the next reporting time?

- What are the major results since the last project review?
- What is the ratio of percent completion to budget depletion?
- Is the project plan still the same?
- What resources are needed for the next stage of the project?

A formal structured documentation of what questions to ask can guide the project auditor in carrying out project audits in a consistent manner. The availability of standard questions makes it unnecessary for the auditor to guess or ignore certain factors that may be crucial for project control.

MEASUREMENT SCALES FOR PROJECT COORDINATION

Project coordination for control purposes requires data collection, measurement, and analysis. Effective coordination is predicated on facts and data. In project management, the manager will encounter different types of measurement scales depending on the particular items to be controlled. It may be necessary to collect data on project schedules, costs, performance levels, problems, and so on. The different types of data measurement scales that are applicable are presented below:

NOMINAL SCALE OF MEASUREMENT

Nominal scale is the lowest level of measurement scales. It classifies items into categories. The categories are mutually exclusive and collectively exhaustive. That is, the categories do not overlap and they cover all possible categories of the characteristics being observed. For example, in the analysis of the critical path in a project network, each job is classified as either critical or not critical. Gender, type of industry, job classification, and color are some examples of measurements on a nominal scale.

ORDINAL SCALE OF MEASUREMENT

Ordinal scale is distinguished from a nominal scale by the property of order among the categories. An example is the process of prioritizing project tasks for resource allocation. We know that first is above second, but we do not know how far above. Similarly, we know that better is preferred to good, but we do not know by how much. In quality control, the ABC classification of items based on the Pareto distribution is an example of a measurement on an ordinal scale.

INTERVAL SCALE OF MEASUREMENT

Interval scale is distinguished from an ordinal scale by having equal intervals between the units of measure. The assignment of priority ratings to project objectives on a scale of 0 to 10 is an example of a measurement on an interval scale. Even though an objective may have a priority rating of zero, it does not mean that the objective has absolutely no significance to the project team. Similarly, the scoring of zero on an examination does not imply that a student knows absolutely nothing about the materials covered by the examination. Temperature is a good example of an item that is measured on an interval scale. Even though there is a zero point on the temperature scale, it is an arbitrary relative measure. Other examples of interval scale are IQ measurements and aptitude ratings.

RATIO SCALE OF MEASUREMENT

Ratio scale has the same properties of an interval scale, but with a true zero point. For example, an estimate of zero time units for the duration of a task is a ratio scale measurement. Other examples of items measured on a ratio scale are cost, time, volume, length, height, weight, and inventory level. Many of the items measured in a project management environment will be on a ratio scale.

Another important aspect of data analysis for project control involves the classification scheme used. Most projects will have both quantitative and qualitative data. Quantitative data require that we describe the characteristics of the items being studied numerically. Qualitative data, on the other hand, are associated with object attributes that are not measured numerically. Most items measured on the nominal and ordinal scales will normally be classified into the qualitative data category while those measured on the interval and ratio scales will normally be classified into the quantitative data category. The implication for project control is that qualitative data can lead to bias in the control mechanism because qualitative data are subject to the personal views and interpretations of the person using the data. As much as possible, data for project control should be based on a quantitative measurement.

DATA ANALYSIS FOR PROJECT COORDINATION

Project data may be obtained from several sources. Some potential sources are:

- Formal reports
- Interviews and surveys
- Regular project meetings
- Personnel time cards or work schedules

The timing of data is also very important for project control purposes. The contents, level of detail, and frequency of data can affect the control process. An important aspect of project management is the determination of the data required to generate the information needed for project control. The function of keeping track of the vast quantity of rapidly changing and interrelated data about project attributes can be very complicated. The fundamental steps involved in project data requirement analysis are as follows:

- Data collection
- Data processing to generate information
- Decision making
- Implementation of action

Data collection constitutes a crucial point in project control. Project decisions require information and information requires accurate data. Effective management requires proper information management. Information can be defined as the resource that facilitates effective project control. Data is processed to generate information. Information is analyzed by the decision maker to make the required decisions. Good decisions are based on appropriate information, which in turn is based on reliable data. Data analysis for project control may involve the following functions:

- Organizing and printing computer-generated information in a form usable for decision makers
- Integrating diverse hardware and software to communicate in the same project environment
- Using voice-activated computerized data analysis to expedite data processing and to reduce paperwork
- Incorporating new technologies such as expert systems into data analysis
- Preserving the integrity of data as they are transformed from one form to another to generate different types of information
- Incorporating flexibility and sharing options into project communication networks.

A comprehensive documentation of project data requirements should be developed. If data is properly documented, the chances for misuse, misinterpretation, mismanagement, or mishandling will be minimized. Data is needed at every stage in the life cycle of a project from the problem identification stage through the project phase-out stage. The various stages of data requirements in project management include the following:

1. Data related to problem identification
2. Data resulting from an initial study of the problem
3. Data on personnel and resource availability
4. Project planning data
5. Data on project initiation
6. Project schedule data
7. Project implementation data
8. Data on project tracking
9. Performance measurement data
10. Project phase-out data

The components of the documentation of data requirements should address the following items:

DATA SUMMARY

Data summary is a general summary of the information and decision for which the data is required as well as the form in which the data should be prepared. The summary indicates the impact of the data requirements on the organizational goals.

DATA PROCESSING ENVIRONMENT

Processing environment identifies the project for which the data is required, the user personnel, and the computer system to be used in processing the data. It refers to the project request or authorization and relationship to other projects, and specifies the expected data communication needs and mode of transmission.

DATA POLICIES AND PROCEDURES

Data handling policies and procedures describe policies governing data handling, storage, and modification and the specific procedures for implementing changes to the data. Additionally, they provide instructions for data collection and organization.

STATIC DATA

Static data describes that portion of the data that is used mainly for reference purposes and is rarely updated.

DYNAMIC DATA

Dynamic data describes that portion of the data that is frequently updated based on the prevailing circumstances in the organization.

DATA FREQUENCY

Frequency of data update specifies the expected frequency of data change for the dynamic portion of the data, for example, quarterly. This data change frequency should be described in relation to the frequency of processing.

INPUT INTERFACE

Data input interface specifies the mechanism through which the data is entered. Data may be entered directly by the user through the computer keyboard, mouse, or light pen. Data may also be entered indirectly by retrieving them from an existing database.

DATA CONSTRAINTS

Data constraints refer to the limitations on the data requirements. Constraints may be procedural, such as based on corporate policy; technical, such as based on computer limitations; or imposed, such as based on conflicting project requirement.

DATA COMPATIBILITY

Data compatibility analysis involves ensuring that data collected for present project needs will be compatible with future needs.

DATA CONTINGENCY

Data contingency plan concerns data security measures in case of accidental or deliberate catastrophe affecting hardware, software, or people.

SCHEDULE COORDINATION AND CONTROL

The Gantt charts developed in the scheduling phase of a project can serve as the yard-stick for measuring project progress. Project status should be monitored frequently. Symbols may be used to mark actual activity positions on the Gantt chart. A record should be maintained of the difference between the actual status of an activity and its expected status. This information should be conveyed to the appropriate personnel with a clear indication for the required control actions. The more milestones or control points there are in a project, the easier the control function. The larger number allows for more frequent and distinct avenues for monitoring the schedule. That way, problems can be identified and controlled before they accumulate into more serious problems. However, more control points means higher cost of control.

Schedule variance magnitudes may be plotted on a time scale (e.g., on a daily basis). If the variance continues to get worse, drastic actions may be needed. Temporary deviations without a lasting effect on the project may not be a cause for concern. Some control actions that may be needed for project schedule delays are as follows:

SCHEDULE CONTROL ACTIONS

- Job redesign
- Productivity improvement
- Revision of project scope
- Revision of project master plan
- Expediting or activity crashing
- Elimination of unnecessary activities
- Reevaluation of milestones or due dates
- Revision of time estimates for pending activities

PERFORMANCE COORDINATION AND CONTROL

Many project performance problems may not surface until after a project has been completed. This makes performance control very difficult. Effort should be made to measure all the interim factors that may influence final project performance. After-the-fact performance measurements are typically not effective for project control. Some of the performance problems may be indicated by time and cost deviations. So, when project time and cost have problems, an analysis of how the problems may affect performance should be made. Since project performance requirements usually relate to the performance of the end products, controlling performance problems may necessitate altering product specifications. Performance analysis will involve checking key elements of the product such as those discussed below:

1. **Scope**
   Is the scope reasonable based on the project environment?
   Can the required output be achieved with the available resources?

2. **Documentation**

Is the requirement specification accurate?

Are statements clearly understood?

3. **Requirement**

Is the technical basis for the specification sound?

Are requirements properly organized?

What are the interactions between specific requirements?

How does the prototype perform?

Is the raw material appropriate?

What is a reasonable level of reliability?

What is the durability of the product?

What are the maintainability characteristics?

Is the product compatible with other products?

Are the physical characteristics satisfactory?

4. **Quality Assurance**

Who is responsible for inspection?

What are the inspection policies and methods?

What actions are needed for nonconforming units?

5. **Function**

Is the product usable as designed?

Can the expected use be achieved by other means?

Is there any potential for misusing the product?

Careful evaluation of performance on the basis of the above questions throughout the life cycle of a project should help identify problems early so that control actions may be initiated to forestall greater problems later.

CONTINUOUS PERFORMANCE IMPROVEMENT

Continuous performance improvement (CPI) is an approach to obtaining a steady flow of improvement in a project. This is essentially coordination of efforts toward a common project goal. The approach is based on the concept of continuous process improvement, which is used extensively in quality management functions. The iterative coordination processes in project management can benefit quite well from the concept of CPI. Continuous performance improvement is a practical method of improving performance in business, management, or technical processes. The approach is based on the following key points:

- Early detention of problems
- Incremental improvement steps
- Project-wide adoption of the CPI concept
- Comprehensive evaluation of procedures
- Prompt review of methods of improvement
- Prioritization of improvement opportunities
- Establishment of long-term improvement goals
- Continuous implementation of improvement actions

A steering (coordination) committee is typically set up to guide improvement efforts. The typical function of the coordination committee, with respect to performance improvement, includes the following:

- Determination of organizational goals and objectives
- Communication with the personnel
- Team organization
- Administration of the CPI procedures
- Education and guidance for company-wide involvement
- Allocation or recommendation of resource requirements

Continuous process improvement requires coordination across teams. Figure 5.7 presents a graphic depiction of team interfaces for improvement efforts. Continuous coordination involves passing the "baton" from one team to another team and from one project stage to the next. Coordination interfaces are both internal as well as external. Figure 5.8 represents the conventional fluctuating approach to performance improvement. In the figure, the process starts with a certain level of performance. A certain

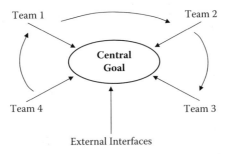

FIGURE 5.7   Team interfaces for project coordination.

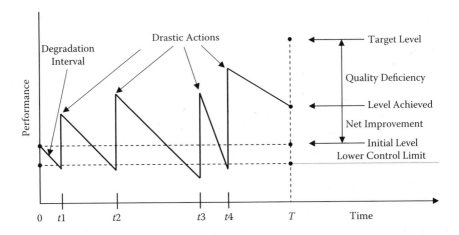

FIGURE 5.8   Erratic approach to project control.

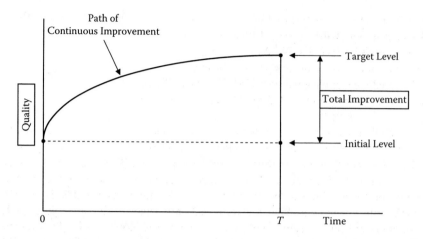

FIGURE 5.9   Progressive approach to project control.

performance level is specified as the target to be achieved by time *T*, the end of the planning horizon. Without proper coordination and control, the performance will gradually degrade until it falls below the lower control limit at time *t*1. At that time, a drastic effort will be needed to raise the performance level. If neglected once again, the performance will go through another gradual decline until it again falls below the lower control limit at time *t*2. Again, a costly drastic effort will be needed to improve the performance. This cycle of degradation-innovation may be repeated several times before time *T* is reached, at which time a final attempt will be needed to suddenly raise the performance to the target level. But unfortunately, it may be too late to achieve the target performance.

There are many disadvantages to the conventional fluctuating approach to improvement. These include the following:

1. High cost of implementation
2. Need for drastic control actions
3. Potential loss of project support
4. Adverse effect on personnel morale
5. Frequent disruption of the project
6. Too much focus on short-term benefits
7. Need for frequent and strict monitoring
8. Opportunity cost during the degradation phase

Figure 5.9 represents the approach of continuous process improvement (CPI). In the figure, the process starts with the same initial quality level, and it is continuously improved in a gradual pursuit of the target performance level. As opportunities to improve occur, they are immediately coordinated and implemented. The rate of improvement is not necessarily constant over the planning horizon. Hence, the path of improvement is curvilinear rather than strictly linear. The important aspect of CPI is that each subsequent performance level is at least as good as the one preceding it. The major advantages of CPI include the following:

1. Better client satisfaction
2. Clear expression of project expectations
3. Consistent pace with available technology
4. Lower cost of achieving project objectives
5. Conducive environment for personnel involvement
6. Dedication to higher quality products and services

A concept similar to that of CPI is the continuous measurable improvement (CMI). Continuous measurable improvement is a process through which employees are given the authority to determine how best their jobs can be performed, measured, and coordinated. Since the employees are continually in contact with the job, they have the best view of the performance process. The employees can identify the most reliable criteria for measuring the improvements achieved in the project. Under CMI, employees are directly involved in designing the job functions. For example, instead of just bringing in external experts to design a new production line, CMI requires that management get the people (employees) who are going to be using the line involved in the design process. This provides valuable employee insights into the design mechanism and paves the way for the success of the design as a customer-centric design.

COST, COORDINATION, AND CONTROL

Figure 5.10 shows a plot of cost versus time for projected cost and actual cost. The plot permits a quick identification of the points at which cost overruns occur in a project. In order to close the cost-benefit gaps, coordination across functions must occur.

Figure 5.11 represents a case of periodic monitoring of project progress. Cost is monitored and recorded on a periodic basis (e.g., monthly). If cost is monitored on a more frequent basis (e.g., daily), then we may be able to have more rigid control structure. Of course, one will need to decide whether the additional time needed for frequent monitoring is justified by the extra level of control provided. The control limits may be calculated with the same procedures used for X-bar and R charts in

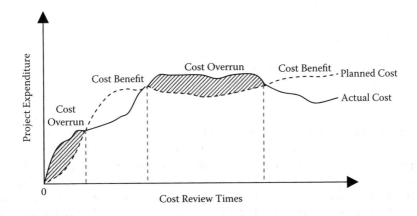

FIGURE 5.10   Plot of resolution of cost and benefit.

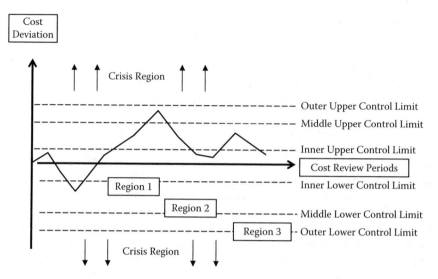

FIGURE 5.11   Periodic review control chart.

quality control, or they may be based on customized project requirements. In addition to drawing control charts for cost, we can also draw control charts for other measures of performance such as task duration, quality, or resource use. In Figure 5.11, cost deviation levels are handled in accordance with this guide.

The region between the inner upper and inner lower control limits: ignore cost deviation.

The region between the middle upper and middle lower control limits: investigate cost deviation later (at analyst's convenience).

Region beyond middle upper and middle lower control limits: investigate cost deviation immediately.

Region beyond outer upper and outer lower control limits (crisis region): investigate cost deviations immediately, high priority.

Figure 5.12 shows a control chart for cumulative cost. The control limits on the chart are indexed to the percentage of the project that is complete. At each percentage completion point, there is a control limit that the cumulative project cost is not expected to exceed. A review of the control chart shows that the cumulative cost is out of control at the 10%, 30%, 40%, 50%, 60%, and 80% completion points. Thus, the indication is that control actions should be instituted right from the 10% completion point. If no control action is taken, the cumulative cost may continue to be out of control and eventually exceed the budget limit by the time the project is finished.

The information obtained from the project monitoring capabilities of project management software can be transformed into meaningful charts that can quickly identify when control actions are needed. A control chart can provide information about resource over-allocation as well as unusually slow progress of work.

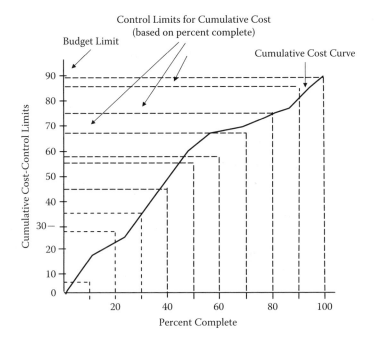

FIGURE 5.12   Control chart for cumulative cost.

Several aspects of a project can contribute to the overall cost of the project. These aspects must be carefully tracked during the project to determine when control actions may be needed. Some of the important cost aspects of a project are presented below:

Cost estimation approach
Cost accounting practices
Project cash flow management
Company cash flow status
Direct labor costing
Overhead rate costing
Incentives, penalties, and bonuses
Overtime payments

The process of controlling project cost covers several key issues that management must coordinate throughout the organization. These include the following:

1. Proper planning of the project to justify the basis for cost elements
2. Reliable estimation of time, resources, and cost
3. Clear communication of project requirements, constraints, and available resources
4. Sustained cooperation of project personnel
5. Good coordination of project functions
6. Consistent policy for project expenditures
7. Timely tracking and reporting of time, materials, and labor transactions

8. Periodic review of project progress
9. Revision of project schedule to adapt to prevailing project scenarios
10. Evaluation of budget depletion versus project progress

These items must be evaluated as an integrated control effort rather than as individual functions. The interactions between the various actions needed may be so unpredictable that the success achieved on one side may be masked by failure on another side. Such uncoordinated analysis makes cost control very difficult. The project managers must be alert and persistent in the cost coordination and monitoring function.

Some government agencies have developed cost control techniques aimed at managing large projects that are typical in government contracts. The cost and schedule control system is based on a work breakdown structure (WBS), and it can quantitatively measure project performance at a particular point in a project. Another useful cost control technique is the accomplishment cost procedure (ACP). This is a simple approach for relating resources allocated to actual work accomplished. It presents costs based on scheduled accomplishments rather than as a function of time. In order to determine the progress of an individual effort with respect to cost, the cost/progress relationship in the project plan is compared to the cost/progress relationship actually achieved. The major aspect of the ACP technique is that it is not biased against high costs. It gives proper credit to high costs as long as comparable project progress is maintained.

INFORMATION FOR PROJECT CONTROL

As the complexity of systems increases, the information requirements increase. Project management has become very essential in many organizations because it offers a systematic approach to information exchange for enterprisewide coordination. Technical projects, in particular, require a well-coordinated communication system that can quickly reveal the status of each activity. Reports on individual project elements must be tied together in a logical manner to facilitate managerial control. The project manager must have prompt access to individual activity status as well as the status of the overall project. A critical aspect of this function is the prevailing level of communication, cooperation, and coordination in the project. The project management information system (PMIS) has evolved as the solution to the problem of monitoring, organizing, storing, and disseminating project information. Many commercial computer programs have been developed for the implementation of PMIS. The basic reporting elements in a PMIS may include the following:

- Financial reports
- Project deliverables
- Current project plan
- Project progress reports
- Material supply schedule
- Client delivery schedule
- Subcontract work and schedule
- Project conference schedule and records
- Graphical project schedule (Gantt chart)

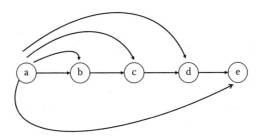

FIGURE 5.13　Feed-forward project coordination.

- Performance requirements evaluation plots
- Time performance plots (plan versus actual)
- Cost performance plots (expected versus actual)

Many standard forms have been developed to facilitate the reporting process. With the availability of computerized systems, manual project information systems are no longer used much in practice. Figure 5.13 reiterates Figure 3.1 to empahsize the importance of communication networking as a feed-forward process for project coordination. The sample project is conceptually moved from stages a, b, c, d, and e. As each stage of the project is reached, information (from tracking and reporting systems) is passed on to the next stage. This is done iteratively until the end goal of the project is achieved.

GUIDELINES FOR TRIPLE C COORDINATION

Coordination across project elements requires that everyone be on the same page. This requires communication and cooperation. In summary, a responsibility matrix should be used to clarify project requirements. The matrix can help resolve questions such as the following:

Who is to do what?
How long will it take?
Who is to inform whom of what?
Whose approval is needed for what?
Who is responsible for which results?
What personnel interfaces are required?
What support is needed from whom and when?

Project integration is what emanates naturally from coordination. The next section discusses the integration of knowledge across the project system so as to achieve synergy of operations.

COORDINATION AND INTEGRATION OF KNOWLEDGE

Project management is a comprehensive endeavor that covers diverse knowledge bases. In addition to the primary body of knowledge defined by the Project Man-

TABLE 5.1

Taxonomy of Expanded Body of Knowledge (E-BOK) for Project Management

E-BOK Qualitative Elements	E-BOK Quantitative Elements
Technology assessment	Systems engineering
Marketing	Operations research
Organizational behavior	Financial analysis
Entrepreneurship	Economic analysis
Intrapreneurship	Decision analysis
Group learning	Cognitive science
Business enterprise	Productivity analysis
Knowledge/technology transfer	Probability and statistics
User interface	Forecasting
Innovation management	Supply network analysis
Strategic planning	Systems simulation

agement Institute in the *Project Management Book of Knowledge*, there are several elements of operation that complement the obvious areas. There are several subtle areas that are just as important. This section introduces the concept of expanded body of knowledge (E-BOK) for project management. E-BOK encompasses all those areas that are important to embrace in light of the emerging complexity and global aspects of contemporary projects. While no single project manager is expected to be well versed in the expanded body of knowledge, it is important to be aware of the elements and to know where to get help when dealing with those elements. A taxonomy of the elements of E-BOK is presented in Table 5.1.

ELEMENTS OF E-BOK

There are several dimensions to achieving an expanded body of knowledge for project management purposes. The dimensions range from issues that are customer-centric to those that are market-driven to those that are organizationally focused. A project system must continually stay on top of most, if not all, of the dimensions by pursuing the guidelines outlined here.

Accountability: Ensuring that everyone knows exactly what they must do in order to meet project goals.

Benchmarking: Conducting programs that objectively compare the organization's project performance with that of competitors.

Business direction: Operating with and consistently communicating a clear objective of the project.

Cash-flow management: Consistently having the ability to meet short-term and long-term project cash-flow obligations.

Competent people: Attracting, continuously developing, and retaining competent and capable employees at all levels of the project organization.

Competitive pricing: Regularly offering products and/or services at a competitive price within and outside the project boundaries.

Cost controls: Consistently keeping project spending within budget allocation levels.

Customer feedback: Actively listening to the customers and using their inputs to improve products and services.

Customer needs: Regularly asking customers what they want, what can be done for them, and how it can be done better.

Customer relationship: Developing and supporting products and/or services based on a long-term project partnership rather than short-term benefits.

Customer responsiveness: Quickly and effectively responding to the needs of project stakeholders.

Customer satisfaction: Listening carefully to customers and addressing any issues that may cause disaffection.

Customer service: Consistently providing service and support to ensure that the customer is satisfied with products and services.

Decision making: Making effective decisions on a timely basis within the constraints of time, cost, and performance expectations.

Development systems: Operating with systems that guide the development of products and services from the initial ideas to the actual offering in the marketplace.

Ethical behavior: Having integrity and being responsible in choosing morally correct actions to achieve goals while understanding the differences and similarities between ethical standards and personal ethics.

Financial stability: Continuously having adequate financial resources to sustain organizational growth or stability.

Integration of work processes: Linking operational and functional relationship of key work processes within a project system.

Life cycle management: Skillfully managing products and services as they pass through the product life cycle stages (introduction, growth, maturity, and decline) in accordance with the iterative steps of project management.

Market image: Communicating a clear and consistent message highlighting the benefits associated with project outputs whether in terms of physical products or services.

Marketing plan: Developing and using a marketing plan that will achieve the objectives of the project.

Measurements: Consistently using accepted performance standards to measure and assess project results.

Motivation: Enrolling and mobilizing all employees to support the project's vision of success.

New product development: Steadily developing and producing new products and services that offer additional value from the project.

Operating plans: Operating with well-coordinated project plans that are revised and updated as necessary.

Operational performance: Operating with work processes to produce products and services at the cost, quality, and schedules required by project stakeholders.

Operational systems: Developing operational systems that allow the organization to grow in an effective manner.

Organizational structure: Operating with a flexible organization that is specifically designed for accomplishing tasks.

Performance reviews: Conducting performance reviews on a formal and regular basis and clearly communicating project status.

Plan of action: Setting goals and executing action plans that ensure the project will stay on a positive course.

Product and service enhancements: Proactively and deliberately improving the characteristics and benefits of all products and services on a continual basis.

Product and service quality: Consistently striving to achieve 100% quality and reliability in all products and services.

Product testing: Actively obtaining reactions and feedback about products and services from project customers and stakeholders.

Productivity improvement: Striving to increase gains from physical and human resources through increased output relative to input.

Profitability: Maximizing financial performance despite the need for other investments such as new product development, training, new facilities, and other infrastructure.

Quality management: Uniformly operating with an all-encompassing philosophy of management based on a vision of quality and customer satisfaction.

Reliable forecasts: Accurately estimating how much of the organization's goods and services must be produced to meet future project needs.

Rewards and recognition: Rewarding people for finding solutions; providing positive recognition for the accomplishment of goals.

Risk taking: Encouraging and taking calculated risks in setting and reaching goals to increase the rate of growth, sales, profits, and competitive position.

Sales leadership: Developing and maintaining a well-trained, highly motivated sales force to disseminate project information promptly.

Sales leads: Regularly developing high-quality sales opportunities as avenues to maximize project acceptance.

Sales organization: Operating with systems that support the sales staff in developing new customers and maintaining existing accounts.

Shareholder value: Continually building increased monetary value for shareholders and project stakeholders and protecting their investments on an ongoing basis.

Skill development: Continuously training employees to develop and maintain competitive skills that will advance the project toward the eventual goal.

Teamwork: Consciously building and using teams of individuals to achieve results, identify new opportunities, and solve problems.

Vision: Consistently communicating what the project organization represents or strives to achieve.

Work environment: Creating a work environment that is conducive for the project team to work and use opportunities for achievement, growth, and accountability.

OBSERVE, ORIENT, DECIDE, ACT (OODA) LOOP APPLICATION TO TRIPLE C

Figure 5.14 shows an observe, orient, decide, act (OODA) loop application within the context of Triple C. The centerpiece of the application is jigsaw model of how tasks within an organization must be observed, oriented, decided, and acted upon. Communication, cooperation, and coordination processes help the organization to achieve the intended end results.

BUILDING PROJECT VALUE AND PERFORMANCE

Any project can benefit from a process of building value and increasing performance. Project management touches every aspect of an organization. It can, thus, be instrumental in building value and performance for the organization. Total systems integration facilitates value streaming across the organization.

Project value is defined as the level of worth, importance, utility, or significance associated with a project in an organization. An input-value-output process transfers inputs (e.g., raw materials, technology, personnel) to desired outputs (e.g., products and services). Typically, value translates to profit for the organization. The merit and justification for a project is a function of several factors as depicted in Figure 5.15.

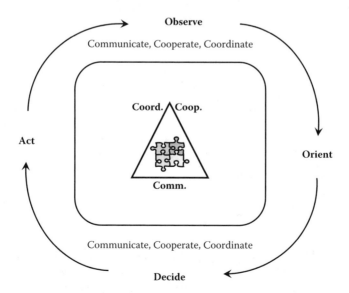

FIGURE 5.14   OODA loop application to Triple C.

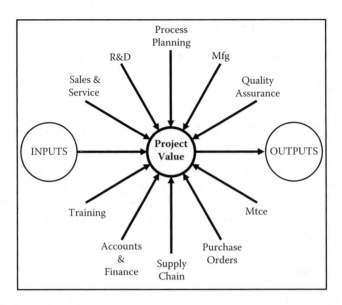

FIGURE 5.15   Project value components.

For instance, sales and service provides a link to the customer through product fol-
low-ups, repairs, and upgrades. Research and development (R&D) provides the basis
for creating, procuring, and implementing new technology for productive application within a project. Process planning, particularly in manufacturing, refers to the
design of detailed work processes. Manufacturing involves creating value for marketing through an organization's production facilities. Quality assurance refers to
the process of achieving and sustaining preservation of product value as perceived
by the customer. Maintenance (Mtce) involves keeping the productive infrastructure
of an organization in peak performance form. Purchase orders transpire through an
organization through supply chain channels to provide an organization with products
and services needed to generate the organization's outputs. Accounts and finance
adds value to the organization by ensuring a balance of credits and debits in the
organization's resources. Training closes the loop of input-value-output by ensuring
that personnel have clear job objectives and fully understand the what-why-who-
how-when aspects of the organization's operations. A proper integration of all these
factors helps to uphold a consistent value stream for a project.

Value stream mapping (VSM) is an important technique used to identify points of
value-added contribution to a project. VSM is normally done by using manual drawing and flow-through analysis of the tasks and steps that constitute a project operation. In manufacturing, VSM is done on the shop floor using a series of sticky notes
on a wall chart. As has been reported in the literature, the hand-drawn value maps
are not very effective communication tools. An analyst can enhance the VSM charts
by incorporating simple communication steps into the process. This requires asking each participant to think through his or her respective operations. We should
recall that Triple C requires full and early communication of why particular tasks

are needed. In the same way, a full and early explanation of why a process needs to be mapped can enhance the VSM technique.

The details collected by the analyst are transcribed into presentation graphics that can be studied as a simulation model of the actual process. A primary purpose of doing value stream analysis is to identify areas of process waste in accordance with lean principles. Waste is generally defined as any activity that does not create value for the end product. The first step in the waste-cleanup process is to take an inventory of all the activities making up a process, with the goal of identifying those that produce value and those that do not. An important tool for this inventory is the value stream map. A value stream map is a representational diagram that describes the sequence of activities a business undertakes to produce a product or family of similar products.

Process steps that are identified as not adding value to the overall project effort are eliminated, thereby conserving limited resources for the more value-adding steps. Lean means the identification and elimination of sources of *waste* in operations. The basic principle of "lean" is to take a close look at the elemental compositions of a process so that non–value-adding elements can be located and eliminated.

KAIZEN ANALYSIS OF A PROJECT

By applying the Japanese concept of *"Kaizen,"* which means "take apart and make better," an organization can redesign its processes to be lean and devoid of excesses. In a mechanical design sense, this can be likened to finite element analysis, which identifies how the component parts of a mechanical system fit together. It is by identifying these basic elements that improvement opportunities can be easily and quickly recognized. It should be recalled that the process of work breakdown structure in project management facilitates the identification of task-level components of an endeavor. Consequently, using a project management approach facilitates the achievement of the objectives of "lean."

VALUE STREAM MAPPING

Value stream mapping describes the activities required to produce a family of products from the beginning to the end. The process involves walking through the process and literally following the product through each step it undergoes. The VSM analyst records every process encountered using a very specific symbols that are used to represent the flow of material and information through the system. The result is a current state map. The symbols help to quickly identify processes that add value and those that do not. Using the current state map as a baseline, the analyst can then develop future state maps. That is, the ideal system designs reduce or eliminate the non–value-added steps. The key to using VSM effectively is its *symbology* and *iconography*.

While VSM was developed primarily for manufacturing and related processes, it can be adapted for project value analysis to track the flow of materials and information through the project life cycle. The communication component of Triple C, in particular, can enhance VSM. Conversely, the process of documenting VSM findings can enhance communication processes in a project.

ORGANIZATIONAL LEARNING

Organizations learn and advance as a collective body. The Triple C model can facilitate organizational interactions that enhance group learning and advancement. Figure 5.16 illustrates organizational learning within the context of project communication, cooperation, and coordination. A Triple C application to organizational coordination and learning has the following elements:

- Diagnose the work culture and social climate of the organization
- Assess employees' readiness for an organizational restructuring to meet project goals
- Develop a strategic communication plan
- Prepare the employees for new challenges brought on by the project.

Figure 5.17 shows an example of a web plot of factor relations for organizational learning. The central factor represents the primary organizational goal, which is

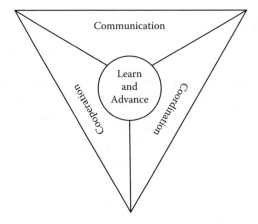

FIGURE 5.16   Organizational advancement through learning.

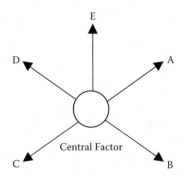

FIGURE 5.17   Web plot for factor relations in organizational learning.

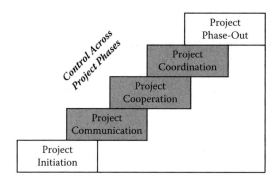

FIGURE 5.18   Triple C steps for organizational learning.

supported by the ambient factors within the organization. Such factors may involve the different business units of the organization working synergistically together in pursuit of the central goal. There is cross-pollination of ideas in between the factors. The angular direction between the factors indicates a measure of factor relationship, proximity, or mutual relevance. The plot can be drawn to increase or decrease the space between two factors to indicate the degree of factor closeness. If one factor fails, a disruption to the flow of work may develop, but some learning effect may still occur from the failure. This creates an experience base that serves the organization better in subsequent endeavors.

Figure 5.18 presents the major steps of project management within the context of organizational advancement from project initiation through project phase-out. It should be noted that each step provides a learning opportunity for the organization. The consistent platform for feed-forwarding the learning from each step to the next is the sequence of communication, cooperation, and coordination that exist in the project. The phase-out of the project itself represents a learning opportunity that is transferred to the next project endeavor. Specific aspects of organizational learning are discussed in the sections that follow.

ORGANIZATIONAL LEADERSHIP BY EXAMPLES

- Leadership system that defines and communicates organizational direction, vision, and major objectives
- Strong leadership team that is not dependent on just one individual
- Leadership that sets goals to improve performance in areas of health and environment protection
- Leadership that sets examples of being good corporate citizens
- Leadership that designs job and organization structures to promote empowerment, efficiency, employee development, and elimination of non–value-added efforts
- Leadership that empowers employees and teams to implement suggestions rather than relying on suggestion systems
- Leadership that eliminates functional departments and layers of management where possible

PROJECT LEADERSHIP

- Senior project leadership involvement in promoting customer focus and performance excellence
- Leadership involvement in promoting customer focus and performance excellence toward how an organization contributes to the community as a corporate citizen
- Leadership approach to recognition programs tailored to individual and group preferences

EMPLOYEE LEARNING THROUGH PARTICIPATION

Here are some pertinent questions to ask in ensuring the employee to learn through participation in project affairs.

- What is done to ensure the workforce is focused and engaged in satisfying customer expectations?
- What is done to encourage employee participation in project and process improvement?
- How do you motivate the workforce to participate in total project management?
- How does an organization ensure proper resources are made available to employees so that they can participate in total project management?
- How are employee contributions recognized and rewarded?
- How is teamwork encouraged throughout the project system?
- How does the organization communicate the effects of future changes to the workforce?

STANDARDS FOR PROJECT MANAGEMENT

Standardization can facilitate project coordination and organizational learning. The growing number of international projects and widespread adoption of project management leads to a significant increase in the number of individuals across the world that need to communicate within and understand the field of project management.

Standards provide a common basis for global commerce. Without standards, product compatibility, customer satisfaction, and production efficiency cannot be achieved. Just as quality cannot be achieved overnight, compliance with standards cannot be accomplished instantaneously. The process must be developed and incorporated into regular operating procedures over a period of time. Standards define the critical elements that must be taken into consideration to produce a high-quality product. Each organization must then develop the best strategy to address the elements. Standards can be formulated under three possible avenues:

1. Regulatory standards
2. Industry consensus standards
3. Contractual standards

All three types of standards are essential for developing widely applicable project management standards. Regulatory standards refer to standards that are imposed by a governing body, such as a government agency. All firms within the jurisdiction of the agency are required to comply with the prevailing regulatory standards. Consensus standards refer to a general and mutual agreement between companies to abide by a set of self-imposed standards. Contractual standards are imposed by the customer based on case-by-case, order-by-order, or project-by-project needs. Most international standards will fall in the category of consensus. Lack of international agreement often leads to trade barriers by nations, industries, and special interest groups.

An international standard for terminology and concepts would prevent many misunderstandings and increase efficiency of project management. Just as in the case of quality standards, unified standards for project management will aid consistent understanding of expectations. Standardization of project management processes under the guidelines of the International Standards Organization will clarify and unify industry-to-industry practices. ISO standards are accepted across the world and carry a level of authority that is recognized beyond those involved with project management professional associations. It is on the basis of this realization that ISO standards for project management are being pursued by cooperating professional organizations. Once those standards become available, project management will reach new implementation heights and be more widely acclaimed.

In summary, project coordination offers the avenue through which organization efforts are brought into fruition. The preceding chapters have provided guidelines for using the approach of the Triple C model to achieve sustainable communication, cooperation, and coordination in project systems. Chapter 7 presents specific case studies of the application of Triple C model to resolve project challenges.

In conclusion, the primary lesson of the Triple C model presented in this book is not to take cooperation for granted. It must be pursued, solicited, and secured explicitly. The process of securing cooperation requires structured communication upfront. It is only after cooperation is in effect that all project efforts can be coordinated.

6 Project Implementation Template

It is helpful to have a model or template that can be adopted for project implementation purposes. This chapter presents a generic project implementation model that can be adapted and modified for specific needs. The comprehensive model is recommended for the essential functions of project planning, scheduling, and control. The generic template can be modified to fit the needs of a specific project. For some, more details will be needed. For others, many of the steps in the template will not be necessary. Each prevailing project scenario will dictate the final structure of the template. Users will customize and tailor the template to their specific or unique needs.

The knowledge areas in PMI's *Project Management Book of Knowledge* (PMBOK) are not presented in a particular sequence of implementations. Thus, a template, such as the one presented in this chapter, is useful for project execution steps. The various PMBOK knowledge areas can be found in the various stages presented in the template. The standard knowledge areas are time, cost, quality, procurement, integration, scope, communication, human resources, and risk. The stages in the template cover one aspect or another within the knowledge areas.

GENERAL TEMPLATE

1. Planning
  I. Specify project background
    a. Define current situation and process
       1. Understand the process
       2. Identify important variables
       3. Quantify variables
    b. Identify areas for improvement
       1. List and explain areas
       2. Study potential strategy for solution
  II. Define unique terminologies relevant to the project
    a. Industry-specific terminologies
    b. Company-specific terminologies
    c. Project-specific terminologies
  III. Define project goal and objectives
    a. Write a mission statement
    b. Solicit inputs and ideas from personnel
    c. Develop statement of work (SOW)

   IV. Establish performance standards
- a. Schedule
- b. Performance
- c. Cost

   V. Conduct formal project feasibility
- a. Determine impact on cost
- b. Determine impact on organization
- c. Determine project deliverables

  VI. Secure management support

2. Organizing

   I. Identify project-management team
- a. Specify project-organization structure
  1. Matrix structure
  2. Formal and informal structures
  3. Justify structure
- b. Specify departments to be involved and key personnel
  1. Purchasing
  2. Materials management
  3. Engineering, design, manufacturing, etc.
- c. Define project-management responsibilities
  1. Select project manager
  2. Write project charter
  3. Establish project policies and procedures

  II. Implement Triple C model
- a. Communication
  1. Determine communication interfaces
  2. Develop communication matrix
- b. Cooperation
  1. Outline cooperation requirements
- c. Coordination
  1. Develop work-breakdown structure
  2. Assign task responsibilities
  3. Develop responsibility chart

3. Scheduling and Resource Allocation

   I. Develop master schedule
- a. Estimate task duration
- b. Identify task-precedence requirements
  1. Technical precedence
  2. Resource-imposed precedence
  3. Procedural precedence
- c. Use analytical models
  1. CPM
  2. PERT
  3. Gantt chart
  4. Optimization models

4. Tracking, Reporting, and Control
    I.   Establish guidelines for tracking, reporting, and control
        a. Define data requirement
            1. Data categories
            2. Data characterization
            3. Measurement scales
        b. Develop data documentation
            1. Data-update requirements
            2. Data-quality control
            3. Establish data-security measures
   II. Categorize control points
        a. Schedule audit
            1. Activity network and Gantt charts
            2. Milestones
            3. Delivery schedule
        b. Performance audit
            1. Employee performance
            2. Product quality
        c. Cost audit
            1. Cost-containment measures
            2. Percentage completed versus budget depletion
  III. Identify implementation process
  IV. Phase out the project
        a. Performance review
        b. Strategy for follow-up projects
        c. Personnel retention and releases
   V. Document project and submit final report

The model above gives a general guideline for project planning, scheduling, and control. The skeleton of the model can be adopted for specific implementation as required for specific projects. Not all projects will need to address all the contents of the model. Customization will always be necessary when implementing the model. The functions and activities encompassed by the project-implementation model do not have to be carried out in a sequential order. Some functions may overlap as illustrated by the chart in Figure 6.1.

SELLING THE PROJECT PLAN

The project plan must be sold throughout an organization. For an industrial development project, selling the project plan may involve dealing with various community groups. For some projects, the plan may need to be sold at the national level. Different levels of detail will be needed when presenting the project to various groups. The higher the level of management, the lower the level of detail. Top-level management will be more interested in the global aspects of the project. For example, when presenting the project to management, it is necessary to specify how the overall

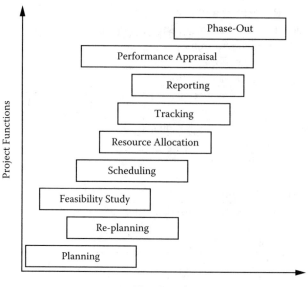

FIGURE 6.1 Gantt chart of project implementation model.

organization will be affected by the project. When presenting the project to the supervisory-level staff, the most important aspect of the project will be the operational level of detail.

At the worker or operator level, the individual will be more concerned about how the project requirements will affect his or her job. The project manager or analyst must be able to accommodate these various levels of detail when presenting the plan to both participants and customers of the project. Regardless of the group being addressed, the project presentation, at a minimum, should cover the essential elements below at the appropriate levels of detail.

- Project background
- Project description
      Goals and objectives
      Expected outcome
- Performance measure
- Conclusions
- Recommendations

TRIPLE C FLOWCHART

Figure 6.2 shows a flowchart model of the implementation of Triple C approach. It shows representative communication, cooperation, and coordination interfaces between units in an organization. Although only four departmental units are illustrated in the flowchart, it can be expanded to include additional units of interest. Figure 6.3 presents Badiru's model of organizational communication using upstream

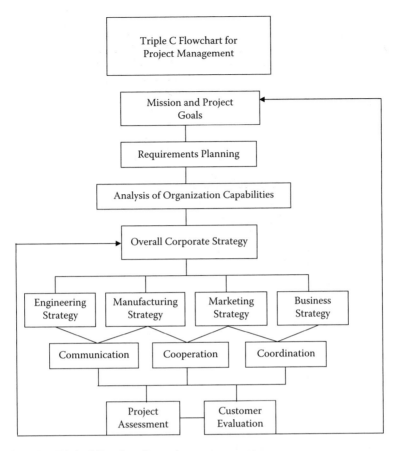

FIGURE 6.2   Triple C flowchart for project management.

and downstream communication channels. This facilitates strategic flow of information and knowledge top-down and bottom-up within an organization.

TRIPLE C IMPLEMENTATION TEMPLATE

COMMUNICATION TEMPLATE

All those who will be affected by the project directly or indirectly, as participants or as beneficiaries, should be informed as appropriate regarding the following questions:

- Why is the project needed?
- What are the expected impacts, potential benefits, costs, and consequences of the project?
- What is the objective and scope of the project?
- Who is in charge of the project?
- When will the project end?
- What personnel contributions are needed?
- How will the project be organized and implemented?

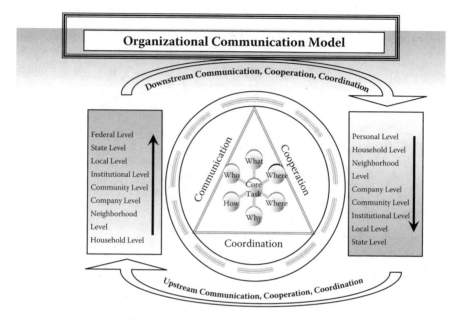

FIGURE 6.3   Triple C organizational communication model. (Copyright Adedeji Badiru, 2006.)

The communication effort can be enhanced if the project team approaches it with the following strategies:

- Exude commitment to the project.
- Demonstrate enthusiasm for the project.
- Pay attention to communication levels of details.
- Involve appropriate organizational hierarchies in the communication.
- Develop a communication responsibility matrix.
- Create multichannel formal and informal communication linkages.
- Identify internal and external communication needs.
- Employ all senses for communication (sight, smell, touch, taste, hearing).

COOPERATION TEMPLATE

Cooperation can be obtained by creating commitment and a sense of involvement among employees. Communication greatly helps to achieve cooperation. A structured approach to seeking cooperation should help identify and explain the following items to the project team:

- The rewards of cooperation and implications of lack thereof
- The level and duration of the cooperative efforts required
- The criticality of cooperation to the project and to the personnel
- The organizational impact of cooperation

COORDINATION TEMPLATE

After successfully initiating communication and cooperation processes, the efforts of the project team must be coordinated. A responsibility chart, which is a matrix consisting of columns of individual or functional departments and rows of required actions, may be developed. This helps to avoid overlooking crucial communication requirements as well as obligations. It also can help to preempt or resolve the following questions:

- Who is responsible for what?
- What personnel interfaces are involved?
- Whose approval is needed for what?
- What support is needed from whom and for what functions?

The Triple C model can be even more effective when used along with other quantitative and qualitative techniques for project management. For example, a software implementation of Triple C can be developed to guide users through the various stages of its application.

COMMUNICATION COMPLEXITY FORMULA

Communication complexity increases with an increase in the number of communication channels. It is one thing to wish to communicate freely, but it is another thing to contend with the increased complexity when more people are involved. The statistical formula of combination can be used to estimate the complexity of communication as a function of the number of communication channels or number of participants. The combination formula is used to calculate the number of possible combinations of r objects from a set of n objects. This is written as:

$$_nC_r = n!/[r!(n - r)!]$$

In the case of communication, for illustration purposes, we assume communication is between two members of a team at a time. That is, combination of 2 from n team members. That is, the number of possible combinations of 2 members out of a team of n people. Thus, the formula for communication complexity reduces to:

$$_nC_2 = n(n - 1)/2.$$

In a similar vein, this book introduces a formula for cooperation complexity based on the statistical concept of permutation. Permutation is the number of possible arrangements of k objects taken from a set of n objects. The permutation formula is written as:

$$_nP_k = n!/(n - k)!$$

Thus, the number of possible permutations of 2 members out of a team of n members is estimated as:

$$_nP_2 = n(n - 1).$$

The permutation formula is used for cooperation because cooperation is bidirectional. Full cooperation requires that if A cooperates with B, then B must cooperate with A. But, A cooperating with B does not necessarily imply B cooperating with A. In notational form, that is:

$$A \rightarrow B \text{ does not necessarily imply } B \rightarrow A.$$

Figure 6.4 shows an example of communication channels in a project network. Figure 6.5 shows the relative plots of communication complexity and cooperation complexity as a function of project team size, n. Complexity increases rapidly as the number of communication participants increases. Coordination complexity is even more exponential since the number of team members increases. Interested readers can derive their own coordination complexity formula based on the standard combination and permutation formulas or other statistical measures. The complexity formulas indicate a need for a more structured approach to implementing the techniques of project management. The templates and generic guidelines presented in this chapter are useful for general management of projects. Each specific project implementation must adapt the guidelines to the prevailing scenario and constraints of a project. The next two chapters present example cases that illustrate how to apply the Triple C model.

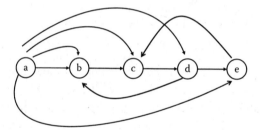

FIGURE 6.4 Example of communication channels in a project network.

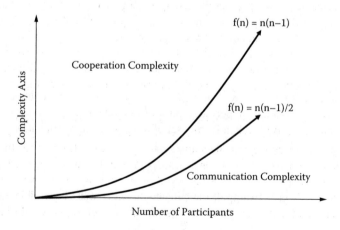

FIGURE 6.5 Plots of communication and cooperation complexities.

7 Triple C Project Case Studies

Example isn't another way to teach, it is the only way to teach.

Albert Einstein

This chapter presents a collection of industry case examples to illustrate typical project problems and how to resolve them using the tools of communication, cooperation, and coordination. The case studies are based on actual industry projects. In some cases, identification details have been rewritten to protect both the innocent and the guilty and preserve the proprietary integrity of those concerned. The case studies present fundamental issues faced by most projects, large and small. The solutions and approaches presented offer a good illustration of how other projects can benefit from the use of the Triple C model. Discussion questions are presented for some of the case studies.

TRIPLE C AT DELTA FAUCET COMPANY

The Triple C approach can be an effective tool when project managers use the process of communication, cooperation, and coordination in an integrated fashion. The Triple C model conveys this idea of integrating these three important characteristics in order to achieve project success. With commitment and the Triple C approach, a project has a much better chance for success than without these key ingredients. Many times, projects fail simply because one of these factors is missing.

This case example is based on a project at Delta Faucet Company in Chickasha, Oklahoma, in the mid 1990s. The project was an enormous undertaking in which the project manager was the overall leader of the project. The project involved the replacement of a Brite Dip system and was justified because of the need to meet an increasing market demand for Delta Faucet products. Because of the complexity of the project, without the utilization of proper communication, cooperation, and coordination, the project would have been doomed to failure. In addition, commitment by all involved was needed in order to achieve the objective of the project and to ensure its success.

Communication was a key ingredient in making all departments aware of their respective responsibilities as well as the progress of the project. The project manager was responsible for conveying the status of the project to all involved. In addition, the project scope, purpose of the project, departmental contributions required, and importance of the project were all conveyed by the manager to all personnel, laterally and vertically. For example, the project manager was responsible for getting approved layouts for the relocation of equipment. Therefore, he would inform the industrial engineer of the need for the layouts. By conducting private departmental meetings as well as general plant meetings, the project manager was able to communicate the required project information to those involved. In addition, regular briefings were scheduled to ensure all involved were aware of the requirements for each individual department as well as the progress of the project. These meetings were done on an informal as well as formal basis.

Cooperation from all involved was also part of the project manager's responsibility. Cooperation is essential in achieving project success because without cooperation from the project personnel, the project would undoubtedly fail. The project manager used communication to disseminate the positive aspects of the project, thereby gaining cooperation for the replacement of the Brite Dip system. In addition, the project manager gained cooperation by making all involved feel important to the overall success of the project. The project manager described each department's responsibility and the criticality of all involved to the success of the project. In addition, the project manager described future spin-off projects that would result from the project, such as a conveyor system to automatically bring parts to work stations. This was of interest to most of the workers. So, they saw a personal benefit in the proposed project. Also, the consequences of a lack of cooperation were conveyed to all involved. The benefits of a cooperative effort to the individual as well as the company were communicated to all involved. Cooperation was needed between many departments for this Brite Dip replacement project because its operation directly impacted many aspects of the company's business. For example, the industrial engineer had to cooperate with the departmental manager and come up with a layout that would satisfy both parties. The project manager's responsibility was to be a middle man and ensure that an agreement was reached.

Coordination of all project efforts was also of critical importance in this project. The project manager was responsible for the coordination requirements. Specifically, determining which department was responsible for time estimates and obtaining the estimates was coordinated by the project manager. In addition, obtaining approvals for certain tasks was also the project manager's responsibility. Also, the project manager ensured that information, support, and consultation interfaces were pursued promptly. Coordination was extremely important on this project because of the need to meet current production demands at the same time as trying to install a new Brite Dip system and tear down the old system.

Using the Triple C approach resulted in achievement of expected outcomes from the project as well as avoidance of any setbacks due to unforeseen events. Making the project proceed in a steady pace rather than having to rush tasks was also a result of proper initial communication, cooperation, and coordination. More specifically, the project would have failed without the use of the Triple C approach. Before embracing the Triple C approach at the author's suggestion, the company faced serious project execution problems, primarily due to a history of uncooperative practices within the company.

Obviously, problems will always arise when any large project is undertaken. The Delta Faucet project was not any different. Such issues as personality conflicts, power conflicts, and priority conflicts will always develop in human interactions.

However, the Triple C model mitigated these conflicts and allowed for the project to move forward and succeed.

CASE STUDY DISCUSSIONS

Consider your own present operating environment and identify a situation involving a change of operations. How would you use the Triple C approach in a way similar to how Delta Faucet used it?

NEW PRODUCT DEVELOPMENT PROGRAM

Project background and plan: The increased demand for widgets in this country and in international markets has provided the impetus for Widgety Manufacturing Company to create a new manufacturing facility to produce and distribute widgets. The current project involves the construction of the production facility. The completion of the project will require the following:

- Capital
- Industrial-grade construction materials, and some subcontracting
- Manufacturing and plant maintenance equipment
- Personnel
  - Nontechnical labor
  - Technical labor
  - Engineering support
  - Managerial and administrative staffing

The plant location should be chosen to facilitate distribution of the product to international markets. This entails finding a location with good access to major shipping, railroad, and highway transportation routes. The production of widgets will meet an increasing international demand for the product, which is forecast to remain high for some time. This will benefit not only the company, but should also improve the country's balance of trade in the international marketplace. The project will require an initial investment of:

- $5 million, which will be used as detailed below:
  - $3 million for materials
  - $1 million for land acquisition and fees
  - $1 million for personnel expenses

The goal of the company is to complete the facility within 18 months of the acceptance of a bid. There are three main objectives within the overall project goal. These are as follows:

- Raising capital for the project
- Acquisition of land, building permits, local regulations, and tax structure
- Plant construction and equipment installation

The capital is to be raised through sales of stock and loan procurement. Clearly, the achievement of this objective will be marked by the production of adequate funds for the project. The acquisition of land for the plant also entails choosing a location, which has a set of local building codes, zoning laws, tax structure, and environmental regulations compatible with the existence of the plant. This objective will be deemed complete when a signed agreement with a community has been reached, all relevant building permits have been acquired, and appropriate tax credits and costs have been addressed.

The third objective is the actual construction of the plant and installation of the industrial machinery. The success of this objective will be measured by the operational ability of the plant. Specifically, it is the ability of the facility to meet all applicable building codes, safety codes, and environmental regulations, as well as the ability for the assembly lines and environmental controls (e.g., heat, air conditioning, plumbing, electricity) to function properly.

Project Approach

Managerial: The managerial approach for the overall project will be the management by objective approach. However, in identified subtasks, local task heads may have latitude in using the management by exception approach, if this is deemed more efficient, particularly from a cost control point of view. In these cases, the local task managers must provide a reasonable, documented rationale for the decision. Accountability will be determined as part of the managerial hierarchy, as well as being based on scheduling concerns, where completion of one task is necessary for completion of others.

Technical: The technical approach to be used in this project relies partly on production technology used in many industries, as well as newer technology, which should improve plant efficiency. The use of newer technology requires that the company identify appropriate personnel for training in the new field in order to facilitate integration of the technology into the plant.

Policies and Procedures

Policy: Policy during project execution is broken into two categories:

* Financial policy
* Physical implementation

Any function requiring the expenditure of funds requires the approval of the manager assigned to the functional group in which the expenditure will take place. Any expenditure exceeding $Z will require the combined approval of the project manager and the manager assigned to fiscal accounting. Physical implementation is a broad category comprising most other policies and their related functions within the project. These policies include setting working hours, length of breaks, worker attire, managerial responsibilities, methods for hiring and terminating, sick leave, and site safety. This list is not exhaustive and will be reassessed and adjusted on a regular basis.

Procedures: Each functional group will have a predetermined budget. Any expenditure within this budget requires written authorization by the group manager prior to disbursement. Any cost overruns must be approved by the project manager after a meeting between the manager, fiscal authority, and group manager.

The policies regarding personnel and site safety will be implemented at the group-manager level. The group manager will ensure that the employees work a full eight-hour day, and that sick leave is only granted when notification is given within eight hours of the start of a shift. Site safety will be based on current Occupational Safety and Health Administration (OSHA) regulations.

Contractual requirement: The communication structure starts with the weekly project review meeting to be attended by all group managers and the project manager. Results of this meeting will be transmitted by the group managers to their respective groups. Each group will have a daily meeting in the morning for the subforeman to address local issues with the group manager. It will be the group manager's job to ensure that significant project concerns from local workers are brought to the attention of the project manager and that universal policies and procedures are determined at the upper management.

Performance specifications will be set by the project manager in the macro level and by group managers at the micro level of management. Performance specifications will include factors such as meeting deadlines, meeting architectural requirements, and meeting cost. Review of performance by each group will be a part of the weekly management meeting.

During a project, large amounts of data are collected while analyzing and reviewing the progress of the project. These data will be kept locally by each group manager, as well as sent to management information service for overall collation and correlation.

Project schedule: The first step in the project is procurement of funds. This will be accomplished within 45 days. When 10% of the desired funding is achieved, preliminary architectural drawings and engineering analyses will be performed. If within 45 days sufficient funding is not obtained, the project will be terminated until it is deemed more financially feasible.

The second phase entails land acquisition. An appropriate site will be chosen that meets the cost, location, and local environmental criteria. This should be accomplished within 30 days. Completion will be defined as that time when all legal, contractual papers have been signed, fund and property rights transferred, and all relevant building permits acquired.

As soon as a site has been identified, during the time the paperwork moves forward for acquisition, a construction firm will be identified. Additionally, engineering support personnel will be recruited. The construction firm will be responsible for hiring the nontechnical and technical personnel. This portion of the project can run partially concurrently with the land acquisition but must be completed within 15 days of the finalization of the land deal. The construction phase of the project will last the remaining 13 months. This phase can be subdivided into several phases:

1. Land preparation (15 days)
2. Necessary in-ground and electric (45 days)
3. Foundation (30 days)

4. Skeletal structure (60 days)
5. Roofing (30 days)
6. Above ground wiring, plumbing, and ducting (60 days)
7. Large equipment installation (60 days)
8. Enclosure (45 days)
9. Remaining equipment installation (30 days)
10. Testing of plant functionality (15 days)

The schedule is derived for sequential completion. However, concurrent work between steps 2 and 3; 5 and 6; 7, 8, and 9 will reduce the overall time. The sequential schedule builds in tolerance for potential delays. Each phase listed above will have a project schedule generated by the functional group manager for that phase.

Resource requirements: The estimated cost of the project is $5 million. Three million is to be spent on materials as follows:

Foundations: $300,000
Electrical Work: $350,000
Plumbing, Ducting: $350,000
Structural Material: $1,000,000
Equipment: $1,000,000

Personnel expenses will be broken down as follows:

Nontechnical labor: $400,000
Technical labor: $200,000
Engineering support: $150,000
Management: $250,000

Part of the $150,000 for engineering support will be used for training specified engineering personnel on new process equipment and design to be implemented in the factory.

Performance measures: The basic measures of performance will consist of the following:

• Completion of task on schedule
• Completion of task at or below budget
• Completed task meets or exceeds all specifications

A weekly audit of expenses for each task will be performed to monitor progress. Completion of the task on schedule is easily verified, but weekly progress reports will be mandatory in order to enable adjustment of scheduling for other tasks if needed. Each task will also be divided into distinct objectives. When each objective is met, an assessment of whether the completed phase meets the specifications of the phase will be made. If the specifications are not met, corrective action will be taken before proceeding. Dividing each task into subtasks should be as compact as possible to ensure that excessive amounts of work are not completed before the identification of sub par components.

Contingency plans: During construction, funds exceeding 15% of the estimated cost will be sought and procured. This will eliminate any delays due to cost overruns. Insurance will be purchased for major natural disasters, as well as for liability and indemnity. Alternate scheduling will be constantly reevaluated to account for delays caused by weather and supply inadequacies.

Tracking, reporting, and auditing: The project manager will be responsible for being knowledgeable about the progress and state of each subgroup. Each week, group managers will supply the project manager with a synopsis report of the current states of his/her group. Each group manager will also submit a more in-depth analysis of the group's operation, in terms of budget, progress, and performance achievements to the management information service. This division will collate and correlate all data and be able to apprise the project manager of the overall state of the project at any time.

CASE STUDY DISCUSSIONS

1. Which organization structure should be used for this product development project? Draw a sketch of the structure.
2. What are some of the international market considerations for this project?
3. Discuss the potential problems with the sick-leave policy.
4. Is the 18-month project schedule realistic? Why or why not?
5. Outline the other considerations, if any, that should be included in the project plan.

B. B. PERRY TECHNOLOGIES

B. B. Perry Technologies is a microcomputer design and manufacturing company that was started in a garage by Bob Perry in 1973. In the early years, Bob ran the company and was largely responsible for making it a huge success. He often worked 20 to 22 hours a day to accomplish that success. The company grew to be an $800 million company in 1980, when Bob sold the company. Substantial stock holdings of the company are presently held by early company employees, who are all of digital electrical engineering background, and who now head the company. Sales of the company peaked at $3 billion in 1984. The company presently employs about 4,500 people.

The market has evolved such that almost all competing companies build to a "de facto" standard (a standard to which Perry products do not adhere), and have gained the majority of the market share. Perry is losing substantial business in both its general-purpose machines and its engineering computer-aided design (CAD) workstation markets. In attempting to resolve its present problems, two camps have arisen. The marketing branch of the business is taking the stance that modifying the Perry product to be compatible with the presently popular standards will solve its present problems, while in the design branch it is widely held that the present market for its products can be expanded through constant redoubling of efforts to make the Perry products technologically innovative and lead the market. To make matters worse, engineers from the manufacturing environment are now arguing that by cutting manufacturing costs, the present products could be "dumped" on the market at extremely low costs.

After attending a project management seminar, the company heads have decided a project management approach could help in the unification of efforts in future product development and have set a course of action to internally train personnel in the marketing, design, and manufacturing groups in concepts of informal project management.

CASE STUDY DISCUSSIONS

1. What are the current company problems?
2. What would you expect the present project management plan to do?
3. What strategy would you suggest in the company's present circumstance?

TRIPLE C MODEL FOR MULTINATIONAL PROJECTS

Bakrie Pipe Industries is one of the major pipe industries in Indonesia. The current plant is located at Bekasi, a suburb area outside the capital city, Jakarta. With the increasing demands in pipe industries, the president of the company instructed the project director to construct and set up a new plant.

The project director is the person who held the highest responsibility for the success of the new plant set up. Based on the organization structure of the company, there are two divisions under his position. These are:

1. The international division: this division is responsible for the out-of-Indonesia business.
2. The development division: this division is responsible for the coordination of all the departments that are involved in the project.

The international division of the project is located in the United States, where the main tasks of the division involve selecting and buying new machines and ensuring good training programs for the use of the new machines. The headquarters of the company is located in Indonesia, where the project director is also located. So, the coordination, communication, and cooperation of the proposed project were difficult.

Owing to the complexities in the coordination, communication, and cooperation between all the divisions and departments in making the project successful, the Triple C model was used extensively throughout the project cycle.

When the project started, the project director held the first general meeting with all the divisions and departments to inform them of the proposed new plant. He informed everyone of the project initiation and asked for his or her full participation when required by the project. Since the international division was not located in Indonesia, the communication was carried out by reporting directly though telephone and fax to the project director in Indonesia. The project director instructed the development division to call a meeting of personnel from the entire department required on this project to discuss further developments.

Cooperation from all the departments was very crucial in achieving the deadline and success of the project. All the departments' personnel that were to be involved had been instructed to direct full attention to the project. This full cooperation helped solve many problems and made the project a success by the deadline.

Coordination of the project was carried out by the development division, which was positioned under the project director. The development division was responsible for coordinating all the departments, assigning tasks to be carried out by each department, and identifying the critical tasks. This ensures that the most important tasks were performed on schedule.

Advantages of Triple C: The Triple C approach made this project a success. Many of the traditional communication problems were overcome or prevented. This was particularly important for the necessary interactions between the international division and the headquarters. The international tasks were integrated successfully. Personnel did not have coordination or communication problems because their tasks had been clearly defined and explained to them up front. Triple C helped eliminate unnecessary costs.

The successful application of Triple C on this international project proved that the approach is effective and efficient. It has paved the way for further use of the approach in future projects of the company.

Implementation considerations: There is often an initial "investment" cost of using Triple C. This involves the time, effort, and/or cost of setting up communication processes, particularly where traditional setups impede personnel communication. However, once the initial obstacles are overcome, the approach provides long-term benefits. Resistance to change is also a factor of concern in the attempt to implement Triple C. Personnel have to be convinced to reorient their task priorities to attend to the requirements of the new project. A lot of time may have to be invested in the attempt to secure the commitment of some department to the proposed project. Many communication efforts may have to be made before full cooperation is achieved. These initial obstacles should not discourage the project efforts.

PROJECT COMMUNICATION AND COORDINATING ACROSS CONTINENTS

The project: The project in this case study is comprised of the engineering, procurement, and construction of seven liquid gas tanks together with their ancillary system and control building. The use and capacity of the tanks are as follow:

1. Three liquefied natural gas (LNG) tanks of 80,000 cubic meters each
2. Two liquefied petroleum gas (LPG) propane tanks of 50,000 cubic meters each
3. Two LPG butane tanks of 50,000 cubic meters each

The ancillary systems that are part of the project include:

1. A propane and butane vapor recovery system
2. A low-pressure flare system
3. An offsite control room

This project is to be built on DAX Island in the Persian Gulf (Middle East), approximately 300 miles from the city of Abu Dhabi (the capital of the United Arab Emirates).

DAX is a small island of two square miles and has a population of 5,000, all of whom work in the only two plants on the island. Owing to the large project size, there will be an additional 3,000 people working and living on the island during the construction phase, which is a considerable increase in the island population. The island has a salty ground and cannot grow any vegetation. Also, there isn't any source for fresh water except for the desalinated seawater; desalination is a very slow process and can be the cause of considerable delays when it comes to concrete pouring.

Each of the seven tanks will consist of two separate and structurally independent liquid containers—a primary inner metallic container and a secondary outer concrete container. Each of the containers will be constructed out of material suitable for the low-temperature liquid.

Each container should be capable of holding the required volume of the stored liquid for an indefinite period without any deterioration of the container or its surroundings. The secondary concrete container should be capable of withstanding the effect of fire exposure from the adjacent tanks and an external impact of 21 tonnes traveling at high speed without loss of structural integrity. Adequate insulation should be provided between the primary and secondary tanks to limit the heat in-leak. All the tanks should be provided with the necessary pumping and piping systems for the receipt, storage, and loading of the LNG and LPG from the nearby plant to the different tankers. A blast proof offsite control room should be provided. All storage and loading operations should be controlled from this control room. The project total cost is estimated at US$600,000,000. The detailed project CPM (critical path method) consists of around 48,000 activities.

Organizational principles: GASC company is the owner of the above project; their main offices are located in Abu Dhabi city. GASC is an operating company that had never managed any engineering or construction project. GASC turned to its mother company, NOCC, for help and signed with them an agreement for the management of this project.

NOCC is the national oil company of the Emirate of Abu Dhabi with main offices in Abu Dhabi city. It is completely owned by the government of Abu Dhabi and plays an extremely important role in the national economy. It is headed by a general manager responsible for the eight different directorates, one of which is the projects directorate. One of the divisions of the projects directorate is the gas projects division, headed by a manager. The gas projects division is selected to manage the above project.

GASC is one of several operating companies whose majority of shares are owned by NOCC. NOCC's share in GASC is 60 percent, and the remaining shares are owned by TOTAC of France, SHELC of Holland, PBC of Great Britain, and MITSC of Japan. While it is considered convenient to have the major owning company manage the project for GASC, such an arrangement has its own drawbacks, mainly the fact that the client company (who should have the final say in its own project) is a subsidiary of the hired management company.

The gas projects division is staffed with a skeleton staff; although very capable, this staff can only oversee the management of the project but cannot perform all the actual engineering, procurement, and construction management. The Houston-based KELLC is selected to provide the services under the direct supervision of the gas projects division. KELLC's scope of work included the basic design of the tanks, the detailed design of all the piping and ancillary systems, the procurement of all free issue materials, and the management of construction. All the engineering and procurement activities are to be performed out of KELLC's regional office in London. KELLC, a well-known process engineering firm, has limited experience with concrete tanks: their selection was conditional on their acceptance to hire the Belgian civil engineering firm TRACT as a consultant to help them with the critical civil engineering problems. NOCC will also hire the U.S.-based consulting firm DMRC to do the soil investigation and testing work.

The construction work is packaged into 15 small contracts and one large (75 percent of the total construction work) contract. All the small contracts were awarded to local construction companies, while the main large contract, which included the tanks, piping and ancillary systems, was awarded to the Chicago-based CBAIC. CBAIC will have to open three new offices: one office in London, next to the KELLC regional office, during the engineering phase; another office in Abu Dhabi city for the construction management; and a third office on DAX island for the construction operations. CBAIC is a reputable tankage contractor. However, its experience in concrete tanks is quite limited. They will have to hire the French civil engineering firm SBC as subcontractors. SBC's experience in low-temperature concrete is limited; they will have to hire the specialized Belgian firm CBC as a consultant.

The project's safety requirements are very high. The safety of those living on such a small island in case of any accident is of a major concern to the owning company. To ensure that the required quality and safety standards are achieved, NOCC will hire the French "Third Party Inspection" company BVC as a consultant.

Since the engineering office is in London, the French and the Belgian engineers are to commute to London as necessary to provide their input to the project. This will continue for the whole engineering phase, which is expected to last two years.

Procurement activities will be handled out of KELLC's London office. Materials and equipment are to be delivered to the construction site. Supplies will be purchased in the open market at the most competitive prices. Steel will be purchased from Japan and Belgium; pipes from Germany, France, and Japan; valves from Sweden and France; pumps from the USA; compressors from Switzerland and Japan; and vessels from Italy. A total of around 600 purchase orders will be issued.

Organizational setup: The organizational setup has to be quite flexible and change in accordance with the project requirements. Since it is expected to be a fast track project, the most active period will be the second half of the engineering phase, which corresponds to the first half of the construction phase. The organizational setup for the project stretches across national boundaries, practices, and regulations. It is, thus, very important that efforts of the project staff be closely coordinated.

CASE STUDY DISCUSSIONS

1. Since this is a project that crosses national boundaries and time zones, communications will, not doubt, be a major problem. It will be desirable to determine a time frame on any given day when all project sites can communicate over the telephone. To address that problem, do the following:

determine a time window on any given day (24 hours) that all the project offices in the following cities or countries can communicate by teleconference: London, Rome, Chicago, Abu Dhabi, Tokyo, Geneva, Bonn, Paris, Sweden, and Belgium. Hint: You may need to make a trip to a library to find out the relative time zones for the countries or cities listed.

2. Carefully go through the narrative of how this project is organized and then develop an organizational chart that conveys that organizational setup. The chart should show the various company names, their responsibilities, their interrelationships, and their locations.

TECH SOUND BANKRUPTCY

Tech Sound, Inc., was a manufacturer of sound equipment for over 50 years. Its product lines included electronic amplifiers and various acoustic speakers. The brand name "TECH 2000" once was the leading brand name and considered to be the Cadillac of the speaker industry.

Tech Sound had two manufacturing facilities, one was located in California and the other was in Oklahoma. The Oklahoma factory production volume made over 70% of the sales even though their corporate headquarters was in California. The product lines of speakers covered industrial and professional units, home stereo, car stereo, and outdoor speakers. All but the industrial and professional speakers were made in Oklahoma. The general techniques and concepts of designing a speaker had not changed for over 30 years. The components needed to assemble a speaker can be divided into two categories:

1. Hardware: such as frame, top plate, bottom plate, pole piece, screws or rivets, and housing (for outdoor speakers only)
2. Software: such as voice coil, spider/diaphragm, corn, corn cap, and gaskets.

These two types of components were manufactured by Tech Sound themselves due to the close and precision tolerance requirements. Tech Sound had large departments of machine shop and coil winding in both factories.

The market of sound equipment is a very competitive one. Products from the United States, Japan, and other foreign countries are all trying to gain bigger market shares. Tech Sound had been experiencing losses in revenue for the last several years, and it had an outstanding debenture, which would mature in another two years. Top management felt a regrouping of the financial structure and a major cost reduction were necessary to keep the company from going under. After several meetings a decision was made to close the factory in California and move the headquarters to Oklahoma. The property in California was sold to pay for the debenture. To have enough floor space for the incoming industrial and professional speaker line and to reduce cost, the final assembly process (assembly of software and hardware to a complete speaker) of the outdoor speakers was moved to Mexico. The required steps of implementing the above decisions are as presented below with responsible departments listed in parentheses:

   I. Planning
     A. Equipment and machinery list (engineering and production)
     B. Production scheduling (production control)
     C. Material inventory list (production control)
     D. Time schedules (project manager)
     E. Transportation requirement (project manager)
     F.   Floor layout (engineering)
     G. Documentation transfer (all)
  II. Physical movement
     A. Ship to new locations (project, engineering, contractor)
     B. Equipment and machinery movement in-house (project engineering, engineering, and contractors)
 III. Training and start-up at new location
     A. Training of assembly personnel
     B. Training on quality control
     C. Training on engineering
     D. Review and evaluation

Tech Sound completed its manufacturing facilities relocation project in 20 months. The production continued at its new location with some problems mainly caused by inexperienced workers (because no employees were willing to be relocated to the new plant). After the learning period was over, the level of production quality and performance went back up to the previous levels. However, one year after the project completion, Sound Tech filed for bankruptcy, and eventually sold out to another company.

CASE STUDY DISCUSSIONS

1. What were the underlying causes of the problems of Sound Tech Company?
2. Were the problems due to poor organizational setups? Should things have been done differently? What would you recommend for a company in an identical situation?

FASTRITE MANUFACTURING COMPANY

In the Fastrite Manufacturing Company (FMC), the engineers are required by the line managers to provide the assembly personnel with manufacturing layouts for every item they manufacture. These manufacturing layouts contain all of the information that the shop needs to build the product. The problem with these documents is the fact that they are always subject to changes due to several factors. Currently, the process to issue the manufacturing layout takes approximately 38 days. The line manager (Diana) feels that this long lead time is not acceptable. Therefore, she called the engineering manager to complain. The engineering manager (Bill) understood the problem and agreed to help.

Bill decided that he would ask John, one of his engineers, to work on this problem. Bill called John into his office to explain to him the situation. John said that he was aware of the problem and that he knew why the problem existed. Bill was pleased with the quick response and wanted to know what the problem was and how he was going to resolve it. John proceeded to explain that the current procedures are uncontrolled and do not contain time constraints: "The process that is associated with issuing manufacturing layouts is far from ideal. Everybody wants to influence the layout in one way or another. In order to complete a manufacturing layout, you are required to obtain nine other engineers' signatures. Each of these nine engineers has a functional interest in the layout. The problem is that there are no time constraints that indicate how long after you receive a layout you should supply your input and pass it on.

Bill then understood what was happening. The layouts were sitting idle on engineers' desks for days before they would do their part. "How are you going to resolve this problem?" Bill asked John. John thought for a minute and decided to develop a PERT chart for the operation. Then he went to each of the nine functional engineers individually to determine the order of precedence and the duration of their activity. After compiling his data, he drew the PERT chart to determine the overall expected duration. He was pleased to find out that the expected duration was only 12 days.

John's next dilemma was what to do with this information in order to decrease the current duration. He decided that the best way was to publish his findings to the functional engineering managers. In doing this he can let the managers convey the information to their engineers, along with instructions to adhere to the time schedule. He reported back to his boss to explain his proposal and get his approval. Bill reviewed the proposal and was thoroughly impressed with the concept and gave John his approval. John's concept worked. The duration was shortened to an average of 11 days. However, John stepped on a few toes by giving the engineering managers the impression that their functional engineers were not doing their jobs expeditiously.

CASE STUDY DISCUSSIONS

1. How could John have introduced the new time constraints to the functional engineers without stepping on their toes?
2. Should Bill be blamed for the animosity that developed?
3. Since the concept worked to the benefit of the organization, should anyone be concerned that some people don't like the way the concept was introduced?

ADVANCED COMMUNICATIONS COMPANY

Advanced Communications is a multinational electronic components and manufacturing corporation. The microwave division (ACM) has a manufacturing plant in Wichita, Kansas. Bill Howard is a failure mode analysis engineer who performs testing on components for ACM's biggest product. The Wave-16 is a short-range communicator used extensively in the military. Currently, ACM has a $400 million contract to produce Wave-16s. The contract will expire at the end of the next year with almost certain renewal if no problems are exhibited by ACM's product.

Four months ago Bill was approached by a product engineer, Al Fleming, about a component that was causing problems both in the factory and in the field. This component, the RB175, is a complex integrated circuit that made up the core of the Wave-16. As Al was becoming more familiar with his product, his testing of the RB175 was becoming more precise. Lately, Al had noticed a trend in the factory of excessive RB175 failures. Since the RB175 was an expensive custom component made at another Advanced Communications facility, Al had collected all the failures and retested them. During retest, 85% of the failed RB175s passed. He tested them again and this time only 35% of the failed RB175s passed. He decided that he needed the equipment and 15 years of experience that Bill could provide. So, he sought Bill's advice.

After two weeks of intense testing under all conditions, Bill and Al came to the conclusion that specific temperature changes and humidity levels caused a specific circuit in the RB175 to react unpredictably. Both of their direct superiors were then informed of the problem. Considering the importance and high cost of the RB175 ($4,500 a piece), a task force was formed to solve the problem. Forming the task force took two weeks itself because of low interest at the RB175 component manufacturing facility. This facility, Advanced Communications Components (ACC) in Denver had never observed this phenomena and felt that ACM was overstepping their boundaries by performing other than pure functional tests. Cooperation was only received after ACM's plant manager used his power and contacted ACC's plant manager.

Once the task force as formed, ACC would not cooperate until they had repeated all the tests Al and Bill had previously performed and documented. This took four more weeks because of poor staffing and a general feeling that a problem did not exist. Once the failure was validated, ACC became cooperative, but ignored complaints of delays. ACC also informed the task force that the circuit that was failing was packaged at their facility, but generated at ACC's Los Angeles facility. Another four weeks were needed to bring the ACC-LA group into the task force and returning to the point that Al and Bill were at several months ago.

Two months later, a solution was yet to be found, so all the R&D people who had been involved in the RB175 were brought on board. Both Al and Bill had been assigned full-time to the task force and were relieved of all other duties. Bill's lab equipment was dedicated to this project only. Meanwhile, the Wave-16 was still being produced on schedule with the hopes that minimal field failures would occur. The customer was not informed of the problem since a new contract was coming up. Any field failures were replaced immediately at no cost to the customer, but a projected loss of $7,000 per replacement in labor and parts was incurred by the ACM task force. One year later, it was decided to redesign the RB175 and the Wave-16 for less cost due to new technology. Advanced Communications presented the new product design, the Superwave-20 to the military as a better product and was awarded the contract. Al and Bill were both transferred to R&D where they could incorporate their design ideas into the new RB180.

Case Study Discussions

1. What management reorganization may be needed to avoid problems of the past in the new contract?
2. What employee communication problems can you identify in the organization of Advanced Communications Company?

METROPLEX MANAGEMENT

"What am I going to do?" Tom asked Dr. Packer, head of the Applied Research Division of Metroplex International. "All the line people assigned to my project are slowing down production and doing other things to cause my project to come in late and way over budget. They are making it necessary for me to pay them for overtime to get a job done that they should not have any trouble whatsoever getting done in the time that I scheduled for it. The customer is calling me daily wanting a progress report on the project. I am getting tired of telling him that we are behind but are try-ing to get the project back on schedule."

Tom Reese is a new project manager on a big, $50 million project for Metroplex, which is a $220 million a year firm. This contract amounts to approximately one fourth of the firm's yearly business, and if it goes well there is a promise of follow-up business of increasing contract values. Tom was hired for the job of project manager specifically for this project because he had a reputation of being a very good project manager that got the job done on schedule, under budget, and with the highest of quality. The executives of the company felt that they needed someone of this caliber for this job. This action angered many of the employees because the historical posi-tion of the company has been one of promoting people from within.

Many of the employees felt that Mike Johnson, the present line manager, should have gotten the new project manager's job. Therefore, the people were doing substan-dard work on the project to ruin Tom's reputation.

"I need your help to get through to Mike Johnson and his people that this is a very important project for the company. We must bring this project in on schedule and not overrun the cost either. They need to understand that they are just hurting themselves by slowing down this project," said Tom.

"Well," said Dr. Packer, "that is your problem. You are the high-powered project manager on this project. You are supposed to be able to work with all the people that you need to handle your project. Isn't that the job of the project manager? You solve the problem."

"Yes, that is the job of the project manager," said Tom, "and I have tried, but Mike Johnson won't listen to me. He says that I have no authority over him and that you, our boss, will have to tell him to do things differently. In the meantime he just keeps right on allowing his people to waste time and money by causing delays in my project, which is making the customer very unhappy. If we don't do well on this contract then our customer may not award us the lucrative follow-up contracts that are being proposed for this project."

"He is absolutely right about the authority level," said Dr. Packer, "but I am still leaving it up to you to get this problem solved, if we lose the follow-up contracts, then heads will roll. Do you understand?"

CASE STUDY DISCUSSIONS

1. Is promotion from within a company a good policy? Why or Why not?
2. Is the claim about sabotaging Tom's reputation accurate?
3. Identify some of the potential problems associated with putting personal interests above company goals.

4. Why is Dr. Packer not cooperating in solving the project problems?
5. Is power play a major factor in this company?

RAPID DEPLOYMENT FORCE

One-star General Meany is the Wing Commander of a Reserve Air Force Base responsible for support of the Regular Air Force in the areas of air cargo and personnel transportation. His organization consists of several Squadrons, which operate independently in various functional areas. The following is a listing of the organizations and their respective functions.

1. *Flying squadron:* Consists of the pilots, navigators, engineers, and load-masters required to fly the aircraft
2. *Maintenance squadron:* Consists of the aircraft mechanics that are responsible for the repair and maintenance of the aircraft
3. *Operations squadron:* Responsible for assigning missions, scheduling aircraft and aircrews, and preparing flight plans to meet wing requirements
4. *Supply squadron:* Provides all the necessary equipment, supplies, and personal clothing to support wing personnel
5. *Civil engineering squadron:* Provides construction resources for maintenance of physical facilities
6. *Personnel squadron:* Responsible for all personnel needs in the areas of military pay, records, benefits, etc.
7. *Hospital squadron:* Responsible for the health care needs of all personnel
8. *Transportation squadron:* Provides the support personnel who prepare cargo for aircraft loading, process personnel for aircraft transportation, and conduct the actual loading/unloading of the aircraft

During day-to-day operations, General Meany's organizations work with a skeletal work force of approximately 400 personnel. One weekend per month, however, this work force swells to over 1500 personnel as civilian reservists assemble for their monthly training. They are permanently assigned to existing organizations to "fully populate" the structure.

One day, General Meany receives an urgent telephone call from Three-star General Bully who informs him that his base has been selected as a participating unit in the National Rapid Deployment Force. The effect of this announcement means that on a 24-hour notice, General Meany would be required to deploy all of his personnel, aircraft, and a detailed listing of equipment and supplies to an unknown destination, with each aircraft lifting off at a specified time, containing a specified combination of personnel, equipment, and supplies.

As the realization of the task enormity began to sink in, General Meany began to ask himself several questions. "How can an air base recall all of its civilian reservists, brief them about their destinations, provide them with advanced per diem pay, inoculate them with the proper serums for the location to which they are to be sent, and get them on the proper aircraft at the proper time for departure? Not only personnel, how can we assemble the specified equipment at each squadron, deliver it to a common assembly area, weigh and inspect it for hazardous cargo, prepare aircraft loading plans, prepare cargo and personnel manifests, and get the cargo loaded on the proper aircraft in time for takeoff, all with only a 24-hour notice?"

Being the good manager that he was, General Meany realized that this was an excellent time for delegation of duties. He recalled that he had sent one of his brightest staff officers to a project management seminar the previous year. Now was the time to put that knowledge to use.

General Meany picked up the phone and called Major Pain, ordering him to report to him right away. When Major Pain arrived, General Meany outlined the project that the base had just been assigned and the questions that he had already anticipated. Major Pain was asked to put together a plan to accommodate the tasking.

CASE STUDY DISCUSSIONS

1. Assume that you are Major Pain and have just completed a project manager course. Outline a sequence of activities required to successfully implement the requested plan.

WYOVAX COMPUTER RELOCATION

This case study describes the results of a two-month study on a project nicknamed WYOVAX. The study is based on a project led by the author in 1995. Although the case study is over a decade old, it still provides a valuable lesson on how to tackle the perplexing problem of installing information technology (IT) facilities. The project scenario is similar to what is faced even in today's IT-oriented organizations. The case study project involves a problem faced by a large natural resources company in an effort to implement the decentralization of computer facilities. The goal was to develop a workable plan and schedule for transferring a VAX 11/780 computer from Oklahoma City to a remote plant located in Wyoming. Even though the class project was modeled after a real problem, hypothetical companies are used in this case study.

The results of the study show that assuming unlimited resources, the project could be completed in 96 working days. With the project beginning on July 23, 1995, and using a five-day workweek, the project could be completed by December 6, 1995. However, resources are limited, so resource leveling was required. The leveling caused the project completion date to be extended by four days, including a Saturday of overtime work, until December 12, 1995. Total project cost including labor, materials, and contractors was estimated to be $165,700.

Project background: An extensive study of the technical computing needs of Synfuels Corporation, a Division of Natural Resources, Inc. (NRI), was completed in March 1995. The study concluded that the current centralized computing resources located at the corporate office in Oklahoma City could not support the expanding needs of Synfuels. A data communications network was being used to connect Synfuels Corporation with the NRI corporate VAX computer system.

The most cost-effective alternative for NRI was to decentralize by transferring a VAX 11/780 processor from Oklahoma City to the remote production site in

Wyoming. This alternative offered the advantages of a decentralized system, which included reduced data communications costs and improved system reliability. In addition, corporate policy supports decentralization and a precedent had been set when the first remotely located computer was installed at Synfuel Corporation's Illinois operation in June 1994.

Top management reviewed the study and agreed on May 1, 1995, that the VAX 11/780 should be moved to Wyoming as soon as was technically feasible. Also, management set a deadline that the system be operational no later than December 9, 1995, in order to meet the scheduled implementation of an integrated equipment maintenance package.

Project organization: A project team consisting of four engineers was formed for the purpose of developing a plan to transfer a VAX 11/780 computer system to Wyoming and make it operational. The goal of the WYOVAX team has been to identify, detail, and schedule the activities required. Resources such as manpower, equipment, and funds were determined. Tasks were broken down along technical and functional lines.

Technical
    Site preparation
    Deinstallation and shipping
    Installation
    Communications
Functional
    Project coordinator
    Project description report
    Project organization report
    Project plan and schedule report
    Periodic progress report
    Presentation to management
    Documentation of project

A schedule of activities for the project was established, which was depicted on a Gantt chart. The engineers independently worked out the activities for each technical area and then met, discussed, and consolidated a master list. A network chart using a researched set of precedence values and durations was prepared and evaluated.

PROJECT DESCRIPTION

1. *Site preparation:* This included the computer room selection, room layout, construction, raised computer flooring, air conditioning, fire protection, electrical wiring, and terminal wiring.
2. *Deinstallation and shipping:* This included the contracting for the removal and freight of the VAX 11/780, 75 KVA power transformer, and automatic circuit breaker. The transfer must be coordinated with site preparation. Also, the 11/780 cannot be removed until its replacement, a VAX 8600, is installed in the Oklahoma City Computer Center and is thoroughly tested.

3. *Installation:* This included the wiring of the isolation transformer, contracting Digital Equipment Corporation to make the system operational, the initial installation of the operating system, applications software, and user data, the purchase of media storage cabinets, furniture, supplies, and the training of the systems manager.
4. *Communications:* A "ring-configuration" communications network is to be established among the processors in Wyoming, Oklahoma, and Illinois. This will entail capacity planning, making an inventory of available equipment on-hand, obtaining authorization for expenditures, purchasing the necessary equipment, and getting it installed. A dedicated long-line to connect Wyoming and Illinois will need to be ordered from AT&T.

Project Plan: A CPM activity on node (AON) network containing 46 activities was constructed. The initial network assumes unlimited resources. The critical path consists of ten activities and has a length of 96 workdays. The activities on the critical path include:

- Procure new VAX 8600 for Oklahoma City
- Interface 8600 with 11/780
- Final acceptance test on 8600
- Deinstall transformer and circuit breaker
- Install transformer and circuit breaker
- Interface fire suppression system and breaker
- Install VAX 11/780 in Wyoming
- Connect communications gear to 11/780
- Load and test operating system and software
- Final installation testing

A detailed analysis was conducted for activities, durations, precedence, early start, late start, early finish, late finish, total slack, and resources for the resource-loading schedule.

Resource loading: The resource requirements for each activity were determined. A total of 13 resources were identified and assigned to the tasks. The resources required and the units available follow:

Resource Types	Units Available
1. Design Engineer	2
2. Carpenter	2
3. Plumber	2
4. Electrician	2
5. Information systems engineer	1
6. Air conditioning contractor	1
7. Access floor contractor	1
8. Fire suppression contractor	1
9. AT&T long lines	1

Resource Types	Units Available
10. DEC deinstallation contractors	1
11. DEC installation contractors	1
12. VAX movers	1
13. Transformer movers	1

A Gantt chart of the project schedule assuming activities can begin at their early start time was constructed. The resource-loading plot for each resource was developed. It was seen that there were not enough numbers of design engineers, IS (information systems) Engineers, carpenters, and electricians to schedule each activity at its early start time. Also, the plan based on early start resulted in a fragmented work schedule for many of the resources. Resource leveling was necessary to correct these problems. The advantage of the resource-loading schedule is that the project can be completed by December 6, 1995, three days before the deadline.

Resource leveling: Resource leveling was required because there were peaks in the manpower required above the level of manpower available. The leveling was accomplished by using the minimum total slack (MTS) heuristic rule, whereby the competing activities with the least total slack were scheduled first. An attempt is made to achieve a uniform work schedule whenever possible.

The resource leveling caused an increase in overall project duration from 96 workdays to 100 workdays. The project length increased because two competing activities, "install isolation transformer" and "interface fire suppression and auto-circuit breaker" were both on the critical path, had been scheduled concurrently, and each required two electricians when only two were available. If two additional electricians could be hired for that period of time, the project duration would not have increased. The resource-leveled Gantt chart showed that no critical path exists for most of the project. This occurred because slack was introduced during project leveling. The resource-leveled schedule has the advantages of meeting the manpower capacities, providing relatively uniform work schedules, and incurring no additional costs. However, the project completion date of December 12, 1995, exceeds the project deadline by three days.

The critical path assuming limited resources consisted of:

- Install Wyoming-to-Illinois communication gear
- Connect communications gear to 11/780
- Load and test operating system and software
- Final installation testing

Incorporating costs with the CPM network can be a valuable tool for planning schedules, evaluating performance, and controlling expenditures during the operation of a project. In the project the total cost was determined by summing the individual costs associated with each activity. The cost estimates were developed from expected actual expenditures for manpower, materials, and contractor services.

TABLE 7.1
Cost Details for WYOVAX Project

	Unit Cost ($)	Cost Basis	Days Worked	Cost ($)
Labor				
Design engineer	200	Day	34	6,800
Carpenter	150	Day	27	4,050
Plumber	175	Day	2	350
Electrician	175	Day	82	14,350
IS engineer	200	Day	81	16,200
Labor subtotal				41,750
Contractor				
Air conditioning	10,000	Fixed	5	10,000
Access flooring	5,000	Fixed	5	5,000
Fire suppression	7,000	Fixed	5	7,000
AT&T	1,000	Fixed	50	1,000
DEC deinstall	4,000	Fixed	2	4,000
DEC install	8,000	Fixed	7	8,000
VAX mover	1,100	Fixed	7	1,100
Transformer mover	300	Fixed	7	300
Contractor subtotal				36,400
Materials				
Site preparation	2,500	Fixed	—	2,500
Hardware	31,900	Fixed	—	31,900
Software	42,290	Fixed	—	42,290
Other	10,860	Fixed	—	10,860
Materials subtotal				87,550
Grand Total				165,700

Labor charges were based on competitive wages plus 25 percent to cover indirect costs such as vacations, holidays, insurance, and so on. The materials specified were deemed necessary to provide a functional environment for the VAX 11/780 and to establish the data communications network. Costs for materials were based on vendor quotes and price lists. The contractors were given general specifications and were asked to submit fixed cost bids for the job.

Total cost of the project was estimated at $165,700, as shown in Table 7.1. Of this cost, 53 percent is made up of materials, 25 percent is direct labor, and the remaining 22 percent is accounted for by contractor fees. A profile of project expenditures versus time (bar chart) over the 21 weeks of the project was developed. The graph is based on the resource-leveling schedule described previously. Of particular significance were peak expenditures that occurred around weeks 15 and 20. An analysis revealed that these peaks were caused by large materials and software purchases over those particular periods.

Results: Assuming the project begins on July 23, 1995, and that resources are unlimited, the project could be completed in 96 working days, by December 6, 1995. The key activity for the "unlimited resource" case is the procurement and installation of the VAX 8600 to replace the VAX 11/780 in Oklahoma City.

Under the assumption that resources are limited, there are insufficient design engineers, IS engineers, builders, and electricians to meet the early start schedule. With resource leveling, a realistic work schedule was achieved. The project completion date was extended by four working days to December 12, 1995.

Case Study Discussions

1. What operational differences exist between IT projects of the 1995 era and IT projects of the present day?
2. How do the differences impact project execution?

SQUEEZING THE EAGLE

Background: American Peripheral Company (APC) is a manufacturer of computer peripheral equipment. APC had dominated their market until recently when Japanese manufacturers acquired the technology to compete effectively with APC. Price erosion forced APC to concentrate on high-performance products, which had the lowest volume, resulting in decreased revenue. Combined with a general industry slump, this has forced APC to sell part of the company and lay off 40% of its employees.

APC implemented project management two years ago to boost productivity and shorten product development cycles. This has been partially successful. Project managers develop schedules with inputs from line managers only to have upper management shorten them by 20 percent to 40 percent to meet perceived market windows. Project managers find that they are four to six months behind schedule when the project begins. Most of APC's employees have been with APC for 10 to 20 years. Though morale is low, many engineers work an extra five to ten hours per week without pay in an attempt to meet their deadlines.

The present: Bob Stephens is the project manager for the Eagle 3, which is currently under development. The project is 11 months into a 15 month schedule and is two months late. The first customer test units are due to be delivered January 2.

Bob has just left a meeting with the general manager and the new products manager (Bob's boss). Bob was told that a major customer must have four test units by Thanksgiving to consider the Eagle 3 for use in his new computer. As on previous projects, Bob was told that getting this contract would make APC "get well" again. The general manager also told Bob that there were no engineers available to reassign to the Eagle 3 project and that the budget would not allow for overtime pay.

Case Study Discussions

1. How should Bob motivate his project team?
2. What options are available for meeting the Thanksgiving deadline?

GENERAL COMPUTER

General Computer is a large-scale manufacturer of state-of-the-art computer equipment, with sales over one billion annually. Their largest factory, in central Arkansas, employs approximately 4,000 people. This factory currently manufactures the 500 ST series of business computers.

In 1977, when the factory started manufacturing the 410 series of computers, the managerial organization was reworked to add a matrix structure project management group to the staff. It consisted of one manager, one assistant manager, and seven technical/engineering personnel. This group made marked improvements in the manufacturing process in the four years from 1977 through 1981.

In 1981, the 410 series was phased out, and the new 510 ST series was introduced. The 510 ST series was much bigger in manufacturing scope and was more complex than the old 410 series. Because of these factors, management decided to add to the project management staff. Instead of adding to the existing project management group, another department was added with another eight nontechnical people. The project management functions were then split, with the old group overseeing the engineering/planning functions and the new group taking the tracking/implementation functions. Both groups reported to the same manager, but different assistant managers.

After an initial adjustment both groups performed well in their respective jobs. After six months, however, the quality of the work began to suffer. Certain jobs were not being completed, with the excuse being given that the offending group thought the other group was supposed to do it. Some jobs were being duplicated, often with conflicting conclusions. Bickering between both groups became more frequent. More and more functional groups began to complain to the project manager. Finally, the project manager called a meeting for both groups. He started the meeting by asking "why are we having these problems?"

CASE STUDY DISCUSSIONS

Assume you are a consultant sitting in at the meeting.

1. What would you expect to hear?
2. Do you feel that the new matrix structure was properly implemented?
3. What are your recommendations?

TURMOIL AT AGV, INC.

Background: AGV Inc. currently employs 55 people. The company was started in 1983 by three fundamental people. The key partners are Mr. White (company president), Mr. Nova (vice president marketing), and Mr. KC (vice president engineering). The charter of the company is to produce a highly specialized automated guided vehicle with a robotic arm mounted on board. This vehicle would be used in cleaning rooms, environmentally unsafe areas, and other specialized circumstances.

After the key members had developed the charter, Mr. White and Mr. KC went to work to financing. Mr. KC took his design and went to work finalizing the specifications. After financing was complete they went to work on their first sale. After

18 months they had a firm order. The total system price was approximately 1 million dollars. Shortly after this sale two additional sales went through. The first project required 18 months to complete. Many difficulties came up but were quickly resolved. When considering the high level of technology used and the fact that they were new in that line of business, the project was considered successful. The system was purchased by a major U.S. manufacturer. The second system sale was to the same U.S. manufacturer. The third system was to a major manufacturer in Japan.

Growth period: During this time the company experienced major growth in size. It reached employment level of 45 people (in 44 months). A setback was the termination of the director of software through mutual agreement. That post has since been filled and seems to be working out. They also have had two additional rounds of financing. The majority of their financing comes from a major U.S. financing company with additional coming from a Dutch financing company.

Current status: Currently, they have 65 employees. They have just taken orders on four additional systems, and work is going on to make major changes to two previous systems. To date they have completed only three systems (after 56 months). The marketing department is now working on 45 potential customers and expect orders to increase by 10 to 30 percent in the next nine months.

The fourth round of financing is underway. It appears to be going satisfactorily, with the following exception. The U.S. investors are very concerned that only three systems have been completed. They noted that many schedules are running behind because of special features for individual customers. They also noted that only a few key people in marketing know what the status of each project is and that these marketing people are the only real contact the customers have with the company. A member of the investment team threatens with the following comment: "I expect to see some major organizational changes to occur before our next investors' meeting in six months."

CASE STUDY DISCUSSIONS

1. What is the real problem?
2. What should be done?

AUTOFAST MANUFACTURING

In July of 1985 Autofast, a large supplier of factory automation and control systems, completed a study that showed that improvements could be made in the procedures used for shipping, staging, and installing their sophisticated systems. Product and marketing managers immediately proposed changing the product delivery strategy.

Company background: Autofast is a large corporation with many diverse product lines. They have recently changed their name and corporate direction after many years of success in a related, but shrinking, industry. The corporation is looking to gain dominance in the factory automation equipment market. The corporation is made up of three primary profit centers: custom products, consumer products, and factory automation products, each with its own vice president. Each of the three business units shares common manufacturing, sales, and service organizations. The factory automation division is responsible for marketing, product development, and product management but depends on corporate divisions for manufacturing, sales, delivery, installation, and service.

Product background: The factory automation systems developed by AUTOFAST are the latest in high technology and use components produced by other AUTOFAST divisions. Terminals and controllers designed by the consumer products division, software written by R&D personnel and manufactured by consumer manufacturing, installation and maintenance documentation produced by the service organization, and hardware from the factory automation division form the total system offered. This requires that shipments to customer sites from various divisions be coordinated so that installation and turnover can be accomplished.

Product delivery with quality (PDQ) project: In the past AUTOFAST enjoyed market dominance and was able to assemble the system at the customer site. Consequently, division delivery fall downs could be absorbed in the installation interval at the site. As their products and markets evolved, the customer became less enamored with this concept and AUTOFAST had to begin staging their shipments at regional staging centers. The individual shipments from each division were held at this center until the complete package was ready for delivery and installation. The implementation of staging had the immediate benefit of eliminating the image problems associated with building the system on customer premises and the visibility of delays, but significantly increased the interval from first component shipment to customer cutover. There was the added problem that the staging centers were warehouses and did not have the proper personnel available to handle the task.

Factory automation product managers immediately recognized the problem and attempted to begin coordinating the shipment of products from the various divisions. Despite product management's best efforts, this attempt failed. In June of 1985 product management completed a study of their product delivery process. This study cited some of the reasons previous strategies did not work and recommended the implementation of the product delivery with quality (PDQ) project. The PDQ project called for moving the problem further upstream in the process where the material logistic problems could be more effectively handled. Specifically, it called for the material from several different division factories to be shipped to the primary equipment factory for staging and then be sent directly to the customer site.

Product management proposed that R&D manage the implementation of the project since it required modifying the drawings to effect changes in the ordering and delivery systems. The R&D organization recognized the seriousness of the problem but felt that product management should spearhead the effort because of the level of interdivision coordination and cooperation required. After months of discussion, R&D management employees realized that product management would not manage the project and conceded to manage it themselves. A functional manager was subsequently assigned to manage the project in addition to his regular jobs. The following text is a chronological history of the events that followed.

- July 1985: product management completed a study of the product delivery system and called for implementation of PDQ.
- December 1985: R&D management assigned John Starky to head project PDQ and dictated that it be completed in nine months.

- January–March 1986: Starky assembled his project team, investigated the nature of the problem, proposed numerous alternatives to solve the problem, and evaluated these alternatives.
- April to June 1985: The project team chose an alternative that would minimize the documentation changes required by R&D and require the manufacturing division to update their shipping procedures, which were antiquated and based on business needs that were no longer valid. The team then planned the implementation of the project, which they estimated would take approximately three months (one month longer than dictated).
- August 1986: Implementation bogged down due to manufacturing objections to the changes in the procedure required.
- September 1986: John Starky's fourth-level manager called a meeting with product management, R&D, and manufacturing to get to the bottom of the delay. He demanded that a workable plan be submitted to him, which would lead to project completion in three months, or he would cancel the project. Starky argued that three months was unreasonable based on the interorganizational negotiations involved. So, he was given four months.
- October 1986: The project team submitted a plan that would lead to completion in four months.
- November 1986: Project implementation began slipping again due to manufacturing resistance, unforeseen logistical problems, faltering support from product management, and fall downs in the documentation group headed by Jack Smith. Jack claimed Starky's plan did not account for the other projects affecting the same drawings as the PDQ project.
- December 1986: Starky's fourth-level manager got involved again and called another meeting with his people (including Smith and Starky), product management, and manufacturing. Starky worked out a joint plan and everyone agreed at the meeting to work through the Christmas holidays in order to meet the original February deadline.
- January 1987: Implementation was going according to schedule! It appeared that success was imminent. Product management organization underwent a shake-up in which the product manager was replaced and the organization reduced in size.
- February 1987: Upper level manufacturing executives became aware of the details of the implementation plan requiring the changing of long-standing policies and called a meeting between themselves and Starky. Starky admirably defended and sold the changes but some open issues remained that required product management input. The project was put on hold by manufacturing until these issues could be resolved at a follow-up meeting.

Starky contacted the new product manager for the required input and requested his presence at the follow-up meeting (after all it was product management's idea in the first place). At the meeting the new product manager acknowledged that the manufacturing concerns were valid and committed to investigate them and reconvene the forum in one week. The product manager researched the issues and found that the original experts were no longer with the company. Faced with this loss of expertise, reduced staff, and legitimate concerns, he drafted a letter to everyone involved with PDQ stating his problem and reluctantly withdrawing support from the project. Without product management support, manufacturing would not participate and R&D had no choice but to cancel the project.

Starky's fourth-level manager was infuriated by the chain of events and had a letter drafted documenting the problems encountered, sending it to all division managers. The letter stated that AUTOFAST cannot survive if they continue to handle projects in this manner and called for a meeting to investigate the reasons for PDQ failure. The letter further stated that over $500,000 and 20 months were devoted to the problem with no results and demanded that this be avoided in the future.

CASE STUDY DISCUSSIONS

1. Construct a flowchart of the events narrated and attempt to sort out the chain of problems.
2. Where or what are the sources of the project problems?
3. How can the demand in the last sentence be met?

PRINTED CIRCUIT BOARD ASSEMBLY

In March 1994, XYZ, a medium-sized company in a northeast region of the United States commissioned a task force to perform an investigation of its printed circuit board (PCB) assembly process. The task force observations indicated that the manual method of inserting electronic components into the printed circuit boards was very difficult and time consuming. The task force recommended that an alternative robotic device assembly method for printed circuit boards should be analyzed and considered.

Company background: XYZ manufactures printed circuit boards (PCBs) for customer orders. The company has 2,500 employees. The PCBs produced are primarily used in small- and medium-range computer systems. The company is looking for ways to reduce the cost of manufacturing PCBs and stay competitive in the world market. The company has six major organizations: marketing, manufacturing, research and development, production control, personnel, and finance. The manufacturing organization has the primary responsibility of justifying and implementing new technologies in the assembly process. About $2.5 million was set aside by the company management for factory automation projects for the 1994/95 fiscal year. The PCB assembly improvement task force described the two possible methods of assembling printed circuit boards as stated here.

Manual method of assembling printed circuit board: The manual method of assembling printed circuit boards was a process in which the production worker manually inserted components such as resistors, diodes, modules, transistors, and capacitors into an empty PCB or raw card. The work area was in an approved electrostatic discharge area that consisted of a grounded workbench, chair, tools, and a light box. The manual assembly process started with the verification of paperwork and parts.

The component placement list (CPL), pick list, routing, part numbers, engineering change level, serial numbers, templates, and any special instructions were checked before assembly. The foil, which was a blueprint of component insertions, was placed on the light box. The foil contained the outline for two boards. The components were manually inserted one at a time, according to the foil. Upon completion of both cards, they were each verified with the CPL and template. If any rework or scrap was to be done it was performed in a manner similar to that of insertion. When the printed circuit boards were completely assembled and verified, they were manually placed into trays.

Robotic device method of assembling printed circuit board: The robotic device method of printed circuit board assembly was a process in which the production worker assembled printed circuits using J-1515 robotic device and controller. The robot was used to insert rectangular chip packages varying from one half inch to one and one half inches onto a partially populated printed circuit board. The work area was in an approved electrostatic discharge area that consisted of J-1515 robotic devices and controller, module insertion heads for one inch and one half-inch modules, machine base, clinch unit, application control, cabinet, industrial program computer, module feeder, manual card feeder, card carriers, and a set of inspection templates.

The robotic method of assembly begins in a similar manner as the manual method. The process begins with the verification of the job paperwork and parts. The paperwork is the same as that used in the manual process with the exception of any special instructions unique to the operation of the robotic device. Once the job paperwork and parts are verified, the operator is ready to begin the setup of the tool. The setup consists of the following steps: ensure that all switches are in the on position as outlined in the operating procedure, obtain the proper work board holder, and load onto the fixture. Once the fixtures are loaded, the operator is ready to load the PCB assembly program into memory in preparation for assembly.

The program contains the insertion pattern and indicates which part number is to be loaded in the proper input channel. The program would then position the X-Y table into proper position and guide the robotic arm to the proper channel to pick up a part and insert the part into the PCB. Upon review of the PCB task force report in June 1994, Mr. James Martin, XYZ plant manager, called a meeting of all managers that reported to him to indicate that the following tasks are yet to be accomplished in the 1994/95 factory automation project:

1. How to justify a $2.5 million expenditure for automating the PCB assembly process.
2. How the comparison between the robotic device method of assembling PCB and the manual method of assembling PCB can be made.
3. Concern about the type of project management techniques that should be used for the automation project.
4. Implementation methodology to be used for the project, if the robotic device method of assembling PCB is selected.
5. How to formulate the overall project logistics, coordination, control and organizational dependencies.

Mr. Martin assigned two of his bright project managers to seek answers to the above questions and report back to him within six months. How can this be achieved?

HUGE ELECTRONICS PROJECT MANAGEMENT

Huge Electronics Background. Huge Electronics Company is a designer and manufacturer of electronics equipment that is sold primarily to government/military customers. Located in the western United States, Huge grew rapidly in the 1970s to become one of the nation's largest government contractors with employees in excess of 50,000. Owing partly to Huge's rapid growth, the company organization chart was constantly in a state of flux. Despite the changes, the engineering divisions remained fairly stable in a classic project management structure. The manufacturing division was structured in a matrix organization because of the large investments in manufacturing equipment necessary. Duplicating these equipment purchases for every project would not be cost effective.

Naturally the project managers in the engineering divisions wielded a great deal of power (if not total power) to set policy and make decisions. The manufacturing project managers did not possess the total authority shared by their engineering counterparts, they did however have a strong say in controlling the destiny of their projects, if not the operating policy of the division. Because of the matrix structure, functional and project managers coexisted at the same level in the management hierarchy, both reporting directly to the division manager. While the power in the division was spread evenly between functional and project management, when push came to shove the project managers got their way more often than not. Probably, the connections that project managers possessed up through the project structure led to the influential edge that seemed to exist.

Make versus buy decisions: As a result of the fast growth experienced by Huge, production capacity could not keep pace with demand in many cases. Some of the company's product designs had to be off-loaded either completely or partially for the production phase of a contract. The question of who should/would make the decision whether to manufacture in-house or off-load a particular product (or portion thereof) was always a point of contention. At least three parties influenced the decision: (1) the manufacturing project manager (MPM), (2) the manufacturing functional managers, and (3) the engineering project manager. Initially a manufacturing project plan is published by the MPM. This plan includes the *make versus buy* plan for components, subassemblies, and final assemblies. Unless the plan is met with resistance from the functional management, the plan will go into action. The manufacturing functional managers will generally attempt to influence the project to have the product assembled in either their shops or not, depending on the capacity and the current and future workloads. The engineering project manager can influence make-buy decisions by the way the products are specified on the drawings to be used for manufacturing. If a design is created that has features that the Huge manufacturing facility is incapable of producing, the manufacturing product manager has no alternative but to have the product fabricated by a firm with the necessary capability.

Project Tiger and the cable shop: The decision faced by the Tiger manufacturing project manager regarding the selection of a production location for the Tiger electronic cables is a dramatic example of the make-buy decisions faced by Huge managers. Below is a description of the cast of characters who attempt to influence the Tiger MPM's decision.

Final assembly project engineer: Wally Carr, with 25 years experience with the company; worked his way up through the ranks; has an inherent distrust for the wire and cable shop due to bad past experience; his advice to MPM: "We should set up our own shop over in the new Tiger final assembly building. That way we can have control over our own destiny. That's what we did on the old stingray project and it worked great. Those cable guys never meet their schedules."

Cable project engineer: Charlie Rain, with five years of experience with the firm, previously was in sales for a small electronic distributor. Unbeknownst to anyone at the time, he has purchased an interest in a local wire and cable subcontractor that specializes in doing overflow work from large prime contractors. His advice, "we should off-load these cables to a local vendor. They are a simple design and we need to concentrate our manufacturing engineering efforts on the more complex designs."

Wire and cable department manager: Richard Treece, who recently took over the wire department and has already shown signs of improving a department that definitely needed some improvement. He is also the direct manager of the wire project engineer who is in favor of off-loading. His advice: "I know the department is near capacity right now, but six months from now when the Tiger project comes down the pike, we will be ready to handle it. We will deliver quality cables to meet your schedule."

You are the project manager. You know how important this project is to both the company and your career.

CASE STUDY DISCUSSIONS

1. Should you go with a department that has been chronically delinquent when the contract has a large incentive/penalty clause for on-time delivery?
2. Can you risk sending out a design to a supplier when the design is yet to be proven?
3. Will you go to your superiors to convince them to allow you to invest the funds necessary to set up a new cable assembly shop?

IMPLEMENTATION OF NEW MRPII INFORMATION SYSTEM

In Air Force Material Command (AFMC), there were a myriad of data systems that didn't communicate with each other. So, engineers were forced to look for one answer in several places. This was costly and time consuming. In 1984, Air Force Logistics Command (AFLC), now AFMC, contracted with Data Systems Corporation to install one system to use for industrial operations and eliminate the multitude of individual systems. The system is called Depot Maintenance Management Information System (DMMIS), pronounced as Dee-Miss, which engineers have come to ridicule as "The Mess." As of 1995, the system still wasn't fully operational, and everyone continues to wonder when it would be. It is of interest to investigate whether a better way exists.

The first problem with DMMIS and its implementation is that it requires that the organization change the way it does business. It is known that the business methods and procedures could use improvement, but that requires another case study. Many engineers feel it is ridiculous to change the way they do business to meet the needs of a data system. Rather, they feel that they need to find or design a data system that would complement the way business is already set up.

Some proponents of DMMIS argue that it is the needed system. DMMIS is driven by a modified commercial system called Manufacturing Resource Planning II (MRPII). MRPII incorporates all of the information that is needed to carry out the organization's business in one huge database. This covers labor standards, material requirements and availability, facilities, schedule, and revenue/cost data. Currently, these information pieces are kept in five or more systems. On this consideration, DMMIS sounds like the right answer to the information management problem. Opponents of the new system present four counter arguments:

1. The system seems to be driven from the top down to the production floor. It could be argued that it should be designed on the floor and driven up. Although, engineers provided data to Data Systems Corporation about their requirements for the new system, they still bought MRPII software and informed the organization to change the way it does business.
2. MRPII is used for manufacturing operations, not repair or maintenance. In the Aircraft and Engine Production Divisions of AFMC, they repair and/ or modify aircraft and engines. MRPII doesn't work well in repair because there is too much variability in the repair process. Manufacturing jobs have set amounts of labor, material, and machine time that result in a completed end item. Repair jobs do that too but they run into additional work on almost every aircraft. This variability can cause them to repair or replace items that they can't plan for. It will cost a lot of money and time to plan all the contingencies into DMMIS. The latest rumors (personnel get very few facts) indicate that DMMIS won't be used in aircraft or engine repair centers, anyway. Thus, it appears that a lot of money is being spent on a system that will have limited use. This situation, buying systems of limited use, is how the organization ended up with the present myriad of data systems. So, it appears they are planning to add another one to the list of "white elephants."
3. The shops have been waiting for the full implementation for over ten years. That is too long to wait for almost anything.
4. With the changes in leadership, The command doesn't appear to be committed to the system anymore.

Specific inadequacies: In the MRPII training courses, the instructors pointed to some inadequacies in the present system that won't allow MRPII to work effectively. These are summarized below:

1. MRPII relies on data accuracy. One data item entered by an employee may be used by 100 other employees. Each person must be held responsible and

accountable for the accuracy of their data inputs. This will require a new era of accountability in the organization.

2. MRPII requires material items to be available at the required time in the required place. To accomplish this, Congress must allow the organization to use sole source purchases to speed up parts purchasing.
3. MRPII requires each piece of material to have a single number assigned to it. Now, material is ordered by stock number. The stock number system can link several parts under one stock number. This situation can cause the organization to get different parts than are ordered. In the high precision work of jet engine repair, suitable substitutes are often not acceptable.
4. MRPII relies on a complete, structured bill of materials (BOM) for each end item. These bills don't exist now, nor is anyone working on them. Further, a bill is needed for each configuration of the C-135 aircraft that they work on. So, 26 similar bills are needed for just C-135s. This seems a monumental task. The estimate floating around is that to construct one of the C-135 Bills would require 500 people dedicated to the task for one year. This sounds colossal, but then, these are big aircraft.
5. As taxpayers, many of the employees are concerned about spending over $350 million on a system that may not work.

CASE STUDY DISCUSSIONS

1. Do a project management analysis of this case study and present your views.
2. Why would Congress not allow sole source purchases? Discuss pros and cons.
3. What are the implications of having such a long lead time for the implementation?
4. Is this organization ready for MRPII? Why or why not?
5. How can "The Mess" be cleaned up?

The case studies in this chapter are presented to illustrate the recurring nature of project problems in business and industry. In spite of the long-standing tools and techniques of project management, the basic problems keep resurfacing. It is the view of this author that simple human-centric approaches are needed to overcome the common problems. Once the human issues are addressed, the technical and software tools of project management can become more effective. In cases where the Triple C approach had been used, remarkable results have been achieved. Thus, it is important to spread the word about how to apply the techniques of the Triple C model to all projects. The next chapter (Chapter 8) presents case presentation slides for applications of Triple C to different projects in the oil and gas industry.

8 Triple C Case Application Presentations

Try it, and you will believe.

Adedeji Badiru, 2008

This chapter presents a selected collection of the application of Triple C model to real-life projects. This collection is designed to illustrate the simplicity and effectiveness of the Triple C approach and the resulting immediate benefits that can be derived from the model. It is through showing that observers can believe the effectiveness of the technique. The case examples are drawn from the author's years of international consulting for multinational organizations, particularly in the oil and gas industry.

CASE PRESENTATION 1: MULTINATIONAL PETROCHEMICAL PROJECT

This project used the Triple C approach to improve operational efficiencies and enhance resource use at a multinational petrochemical project. Immediate benefits were realized from the Triple C model as a result of the improved communication and sustainable cooperation that developed among the personnel on the project. Management saw the benefits and decided to incorporate Triple C training into the general training programs of engineers in the organization.

Application of Triple C
PETROCHEMICAL PROJECT

A CASE STUDY

SLIDE 8.1   Petrochemical Project.

SLIDE 8.2   Petrochemical Project—Mission.

SLIDE 8.3   Petrochemical Project—Capacity.

SLIDE 8.4   Petrochemical Project—Plant Component.

PRODUCTS

- ETHYLENE
- PROPYLENE
- STEAM

SLIDE 8.5   Petrochemical Project—Products.

PROBLEMS

- INADEQUATE SPARE PARTS TO SERVICE RUNNING EQUIPMENT:
  - Lead to epileptic plant operations
- LACK OF ADEQUATE TRAINING AND NECCESSARY EXPOSURE TO MEET THE CURRENT TECHNOGICAL SKILL

SLIDE 8.6   Petrochemical Project—Problems.

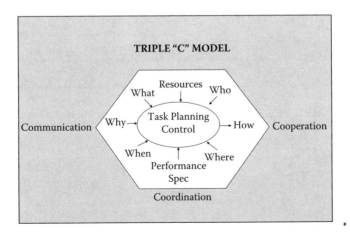

SLIDE 8.7   Petrochemical Project—Triple "C" Model.

TRIPLE "C" ANALYSIS

- <u>HOW</u>: On shutdown or on run
- <u>WHO</u>: Management
- <u>WHERE</u>: Plant
- <u>WHEN</u>: When Required
- <u>WHY</u>: To maintain a continuous running of Plant
- <u>WHAT</u>: Specific parts as specified by the user
- <u>PERFORMANCE SPEC:</u> To achieve high capacity utilization

SLIDE 8.8   Petrochemical Project—Triple "C" Analysis.

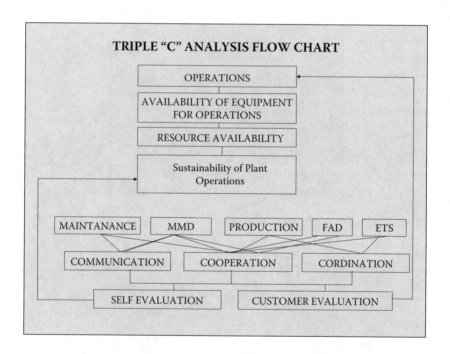

SLIDE 8.9   Petrochemical Project—Triple "C" Analysis Flow Chart.

CASE PRESENTATION 2: IMPROVEMENT OF LOW CAPACITY USE

Low-capacity utilization is a problem that plagues many organizations. Billions of dollars are lost annually due to resource inefficiencies. People are often placed in the wrong jobs. Investment is often misplaced. Wrong tools or skills sets are often applied to organizational challenges. Processes are often dysfunctional. Pieces of capital equipment are often not fully utilized before they reach the end of their useful lives. Then the cycle of low-capacity utilization starts all over again. In spite of the wide recognition of these problems, many organizations still fail to come up with workable solutions. It is the author's contention that the wrong solution approaches are typically applied to these problems. The "big guns" of technical and managerial solutions are often directed at problems whose root causes are simple and centered on human operational flaws. The simplicity of the application of the Triple C model solves most of these problems. The effectiveness of the approach is based on its focus on human-to-human communication, cooperation, and coordination. The case example in this section bears out this fact.

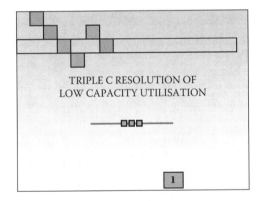

SLIDE 8.10 Improvement of Low Capacity Utilization.

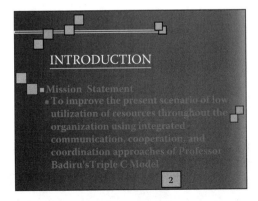

SLIDE 8.11 Improvement of Low Capacity Utilization—Introduction.

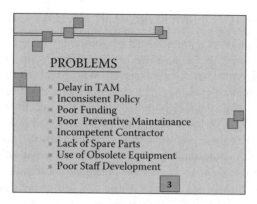

SLIDE 8.12 Improvement of Low Capacity Utilization—Problems.

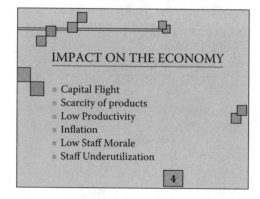

SLIDE 8.13 Improvement of Low Capacity Utilization—Impact on the Economy.

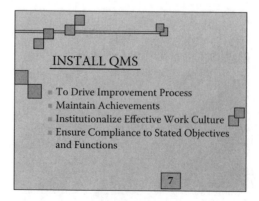

SLIDE 8.14 Improvement of Low Capacity Utilization—Install QMS.

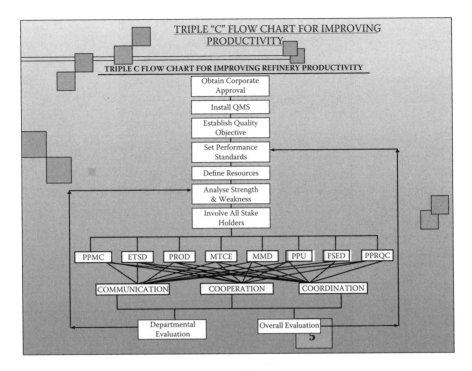

SLIDE 8.15   Improvement of Low Capacity Utilization—Triple "C" Flow Chart for Improving Productivity.

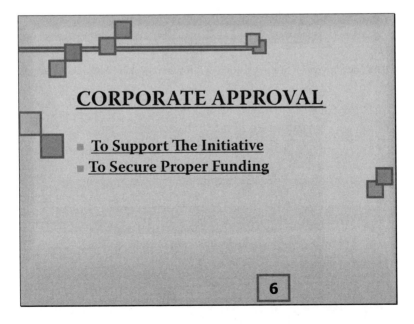

SLIDE 8.16   Improvement of Low Capacity Utilization—Corporate Approval.

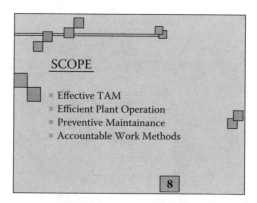

SLIDE 8.17   Improvement of Low Capacity Utilization—Scope.

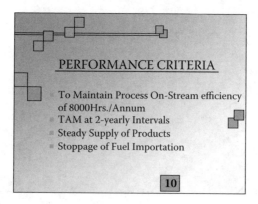

SLIDE 8.18   Improvement of Low Capacity Utilization—Establish Quality Objectives.

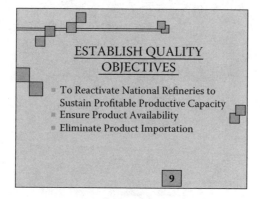

SLIDE 8.19   Improvement of Low Capacity Utilization—Performance Criteria.

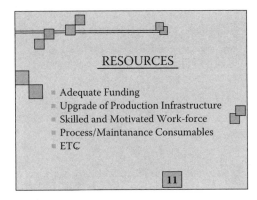

SLIDE 8.20 Improvement of Low Capacity Utilization—Resources.

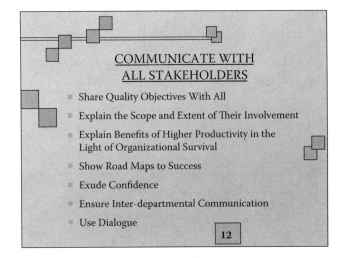

SLIDE 8.21 Improvement of Low Capacity Utilization—Communicate with All Stakeholders.

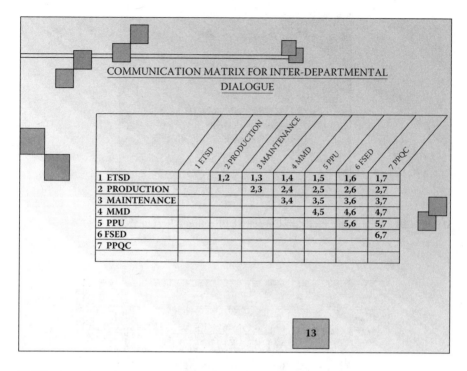

SLIDE 8.22 Improvement of Low Capacity Utilization—Communication Matrix for Inter-Departmental Dialogue.

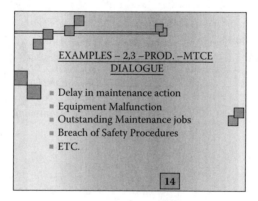

SLIDE 8.23 Improvement of Low Capacity Utilization—Examples-2,3-Prod.-MTCE Dialogue.

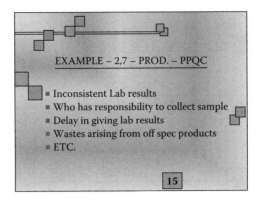

SLIDE 8.24 Improvement of Low Capacity Utilization—Example-2.7-Prod.-PPQC.

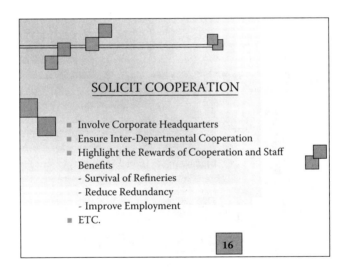

SLIDE 8.25 Improvement of Low Capacity Utilization—Solicit Cooperation.

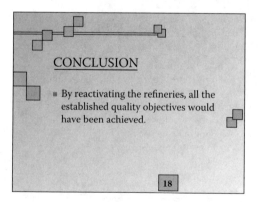

COORDINATION

S/NO	RESPONSIBILITY	Corporate	PPMC	ETSD	PROD	MTCE	MMD	PPU	FSED	PPQC
1	Release of fund	R	I	I	I	I	I	I	I	I
2	Proper Maintenance Policy	I	I	C	C	R	C	C	I	I
3	Carry out TAM as due	R	C	R	R	R	C	R	C	C
4	Supply adequate feedstock	R	R	C	C	I	R	C	I	R
5	Staff training & Dev.	R	R	R	R	R	R	R	R	R
6	Ensure proper product quality	S	S	S	R	I	I	R	I	R
7	Replace obsolete equipment	R	C	R	R	R	R	R	R	R
8	Improve stock for spare parts	S	I	R	R	R	R	R	R	R
9	Monitor progress	R	R	R	R	R	R	R	R	R
10	Institutional improvement	R	R	R	R	R	R	R	R	R

R = Responsible   C = Consult   I = Inform   S = Support

17

SLIDE 8.26 Improvement of Low Capacity Utilization—Coordination.

CONCLUSION

- By reactivating the refineries, all the established quality objectives would have been achieved.

18

SLIDE 8.27 Improvement of Low Capacity Utilization—Conclusion.

CASE PRESENTATION 3: TRIPLE C SYSTEMS INTEGRATION FOR QUALITY IMPROVEMENT

Systems approach to organizational improvement offers a better way to achieve quality goals and objectives. With the increasing shortages of resources, more emphasis should be placed on the sharing of resources. Resource sharing can involve physical equipment, facilities, technical information, and ideas. The Triple C approach to systems integration facilitates the sharing of resources. Integration facilitates the coordination of diverse technical and managerial efforts to enhance organizational functions, reduce cost, improve quality, improve productivity, and increase the utilization of resources. Systems integration ensures that all performance goals are satisfied with a minimum of expenditure of time and resources. It may require the adjustment of functions to permit sharing of resources, development of new policies to accommodate product integration, or realignment of managerial responsibilities. Important factors for integration include the following:

- Unique characteristics of each component in the integrated technologies
- Relative priorities of each component in the integrated technologies
- How the components complement one another
- Physical and data interfaces between the components
- Internal and external factors that may influence the integrated technologies
- How the performance of the integrated system will be measured

For all of the above items, human communication, cooperation, and coordination are necessary. The case example in this section illustrates where and how Triple C can be applied.

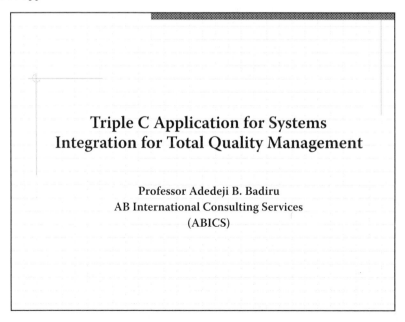

Triple C Application for Systems Integration for Total Quality Management

Professor Adedeji B. Badiru
AB International Consulting Services
(ABICS)

SLIDE 8.28 Triple C Systems Integration.

Everyone's Responsibility for Quality

"Good quality is everybody's responsibility; Bad quality is everybody's fault." Badiru, 1993.

To effectively discharge quality responsibility, strategic **communication, cooperation,** *and* **coordination** *must be in effect throughout the organization. That is, the organization must use the Triple C approach.*

What is quality?
◆ Measure of customer satisfaction
◆ Quality should be defined in terms of user perception

<u>**Use Basis:**</u>

An item that is viewed as having high quality in one application may not be considered to be of high quality in another application.

SLIDE 8.29   Triple C Systems Integration—Everyone's Responsibility for Quality.

Time Basis:

An item that has high quality at one time may not have high quality at another time.

Goal of any organization:
◆ Provide higher quality products and services.

Goal of every worker:
◆ Improve quality and standard of living.

Triple C Systems approach is the best way to achieve the objectives of TQM.

SLIDE 8.30   Triple C Systems Integration—Time Basis.

Systems Definition of Quality (Badiru,1993):

"Quality refers to an equilibrium level of functionality possessed by a product or service based on the producer's capability and the customer's needs."

Producer's Capability

Defined by the aggregate capabilities of:

- Worker's skills, training, and experience
- Machine Capabilities
- Available Production Facilities

SLIDE 8.31   Triple C Systems Integration—Systems Definition of Quality.

Customer's Needs

Defined by a combination of:

- What the customer wants
- What the customer needs
- What the customer favors

Each component of the producer's capability and the customer's needs should be viewed as subsystems of the overall quality management system.

SLIDE 8.32   Triple C Systems Integration—Customer's Needs.

TQM Concept

- ◈ TQM refers to a total commitment to quality
- ◈ "Total" refers to an overall integrated approach:
    - all aspects of quality
    - all the people
    - all the hardware
    - all the software
    - all the organizational resources

- ◈ TQM requires total participation of everyone.
- ◈ *Quality is required from all the people at all times.*

SLIDE 8.33   Triple C Systems Integration—TQM Concept.

An integrated systems approach to total quality management facilitates an awareness of the importance of quality throughout an organization.

A systems approach considers all the interactions necessary between the various elements of the organization.

A system is a collection of interrelated elements working together collectively to achieve a common goal.

The supporting cooperative actions of subsystems in an organization serve to counterbalance the weaknesses at certain points in the organization.

SLIDE 8.34   Triple C Systems Integration—TQM Concept 2.

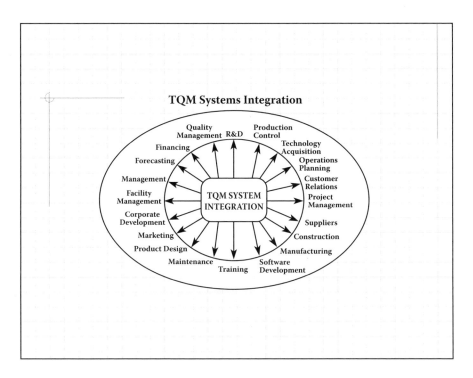

SLIDE 8.35   Triple C Systems Integration—TQM System Integration.

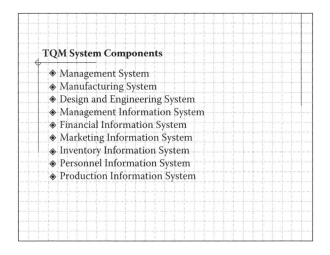

SLIDE 8.36   Triple C Systems Integration—TQM System Components.

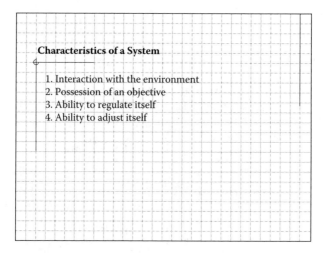

SLIDE 8.37   Triple C Systems Integration—Characteristics of a System.

SLIDE 8.38   Triple C Systems Integration—Characteristics of a System 2.

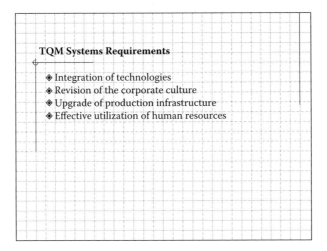

SLIDE 8.39 Triple C Systems Integration—TQM Systems Requirements.

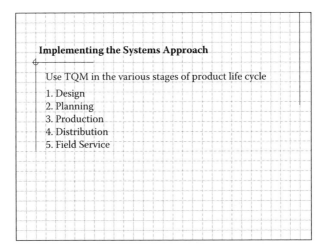

SLIDE 8.40 Triple C Systems Integration—Implementing the Systems Approach.

Management's Responsibility

1. Raise the level of awareness about the implications of low and high quality.
2. Adopt a supportive quality philosophy.
3. Back the philosophy with required resources.
4. Make TQM a requirement throughout the organization.
5. Institute periodic quality reporting requirements.
6. Establish quality liaisons with clients and suppliers.
7. Adopt a flexible perception of systems operations.
8. Play a visible role in quality management.
9. Require that functional managers document how department level quality decisions affect other units of the organization.
10. Understand the limitations of automation in quality management.
11. Management must not only proclaim the need for better quality, but must also commit the necessary resources for it.
12. Investments made for quality today will lead to higher profits in the future.

SLIDE 8.41   Triple C Systems Integration—Management's Responsibility.

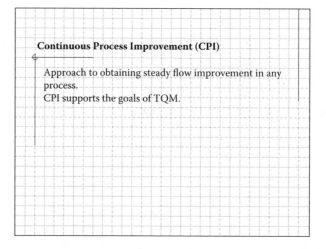

Continuous Process Improvement (CPI)

Approach to obtaining steady flow improvement in any process.
CPI supports the goals of TQM.

SLIDE 8.42   Triple C Systems Integration—Continuous Process Improvement (CPI).

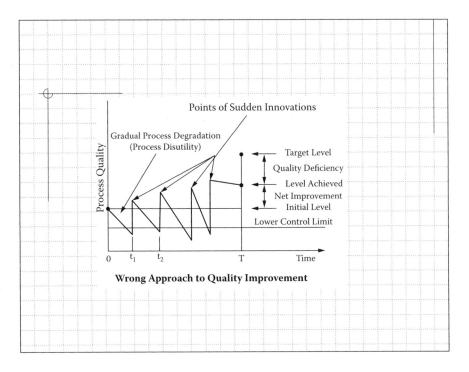

SLIDE 8.43   Triple C Systems Integration—Wrong Approach to Quality Improvement.

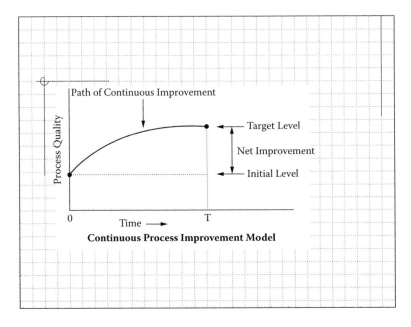

SLIDE 8.44   Triple C Systems Integration—Continuous Process Improvement Model.

Advantages of CPI

- Early detection of problems
- Ability to stay one Generation Ahead (OGA) of the competition
- Lower cost of achieving quality objectives
- Prioritization of improvement opportunities
- Establishment of a conductive decision making team
- Comprehensive evaluation of procedures
- Review of methods of improvement
- Establishment of long-term improvement goals
- Continuous implementation of improvement actions
- Company-wide acceptance of CPI concept
- Better customer satisfaction
- Consistent pace with process technology
- Ability to keep ahead of the competition

SLIDE 8.45   Triple C Systems Integration—Advantages of CPI.

Disadvantages of Not Using CPI

- High cost of implementation
- Frequent disruption of the process
- Too much focus on short-term benefits
- Need for sudden innovations
- Opportunity cost during the degradation phase
- Negative effecton personnel morale
- Loss of customer trust
- Need for frequent and strict monitoring
- Quality control rather than quality management

SLIDE 8.46   Triple C Systems Integration—Disadvantages of Not Using CPI.

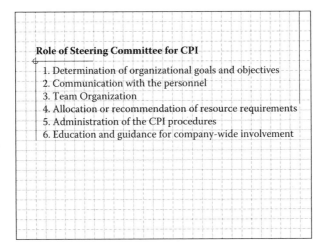

SLIDE 8.47 Triple C Systems Integration—Role of Steering Committee for CPI.

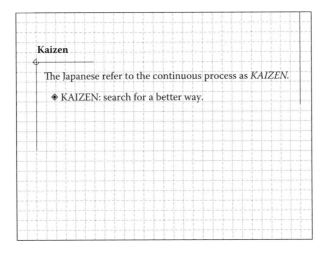

SLIDE 8.48 Triple C Systems Integration—Kaizen.

Continuous Measurable Improvement

◆ Give workers authority to determine how best their
  jobs can be done
◆ Employees are continually in contact with the job
   ▪ Have the best view of performance of the process
   ▪ Have the most reliable criteria for measuring the
     improvements achieved
   ▪ Should be involved in designing the job functions

SLIDE 8.49   Triple C Systems Integration—Continuous Measurable Improvement.

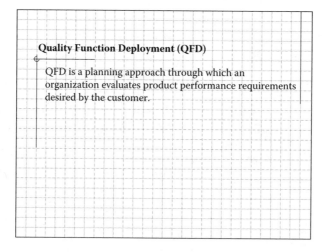

Quality Function Deployment (QFD)

QFD is a planning approach through which an
organization evaluates product performance requirements
desired by the customer.

SLIDE 8.50   Triple C Systems Integration—Quality Function Deployment (QFD).

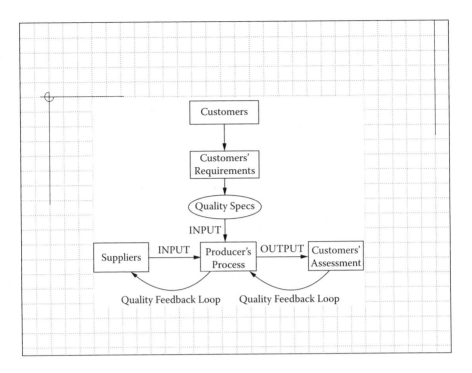

SLIDE 8.51   Triple C Systems Integration—Quality Function Deployment (QFD) 2.

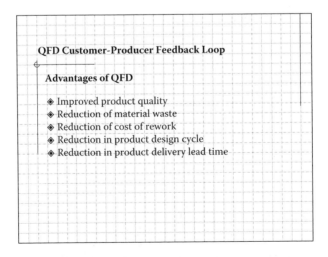

SLIDE 8.52   Triple C Systems Integration—QFD Customer-Producer Feedback Loop.

Implementation of QFD

1. Conduct credible marketing research.
2. Establish **communication** liaison with customer.
3. Make customer aware of capability and limitation or production facility.
4. Establish **cooperation** between design, manufacturing & marketing.
5. Give feedback to the customer on product design and performance.
6. Strive for a trusting relationship with the customer.
7. Institute effective **coordination** points throughout the process.

SLIDE 8.53   Triple C Systems Integration—Implementation of QED.

Deming's 14 Points

1. Create constancy of purpose for improvement of product and service. The aim should be to become competitive and stay in business to provide jobs.
2. Adopt the new philosophy. The time is ripe for changes and new leadership in management approaches.
3. Cease dependence on inspection to achieve quality. Eliminate the need for inspection on a mass basis by building quality into the product in the first place.
4. End the practice of awarding business on the basis of price tag alone. Instead, minimize total cost by using single supplier for any one item to facilitate a long-term relationship of loyalty and trust.
5. Improve constantly and forever the system of production and service in order to improve quality and productivity,and, consequently, decrease costs.
6. Institute training on the job.
7. Institute leadership.The goal of supervision should be to help people and machines and gadgets do a better job. Supervision of management and production workers should be reviewed.

SLIDE 8.54   Triple C Systems Integration—Deming's 14 Points.

8. Drive out fear so that everyone can work effectively
    for the company.
9. Break down barriers between departments. Those
    who are involved in research, design, sales, and
    production must work together as a team.
10. Eliminate slogans, exhortations, and targets for the work
    force requiring zero defects and new levels of productivity.
    Exhortation only creates adversarial relationships. The
    majority of causes of low quality and low productivity can
    be found in the production system rather than within the
    control of the work force.
11. A. Eliminate work quotas on the factory floor.
    B. Eliminate management by objective. Eliminate
    management by numbers and numerical goals.
    Substitute leadership.
12. A. Remove barriers that rob the hourly worker of his/her
    right to pride of workmanship. The responsibility of
    supervisors must be changed from sheer numbers to quality.
    B. Remove barriers that rob people in management and in
    engineering of their right to pride of workmanship. This
    requires the elimination of annual rating or merit systems.
13. Institute a vigorous program of education and
    self-improvement for everyone.
14. Put everybody in the company to work to accomplish the
    transformation. The transformation is everybody's job.

SLIDE 8.55   Triple C Systems Integration—Deming's 14 Points 2.

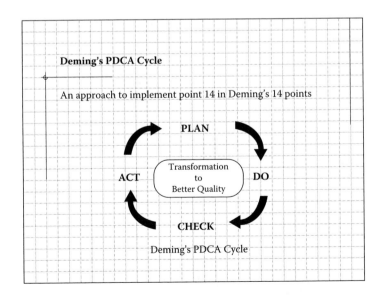

SLIDE 8.56   Triple C Systems Integration—Deming's PDCA Cycle.

Step 1: Plan (Triple C is Applicable Here)

Determination of what is to be achieved and how
it will be achieved:
- What is the specific objective?
- Who are the members of the team?
- When is the plan to take effect?
- When is the plan expected to end?
- What will be needed to carry out the plan?
- What data are already available?
- What data will need to be collected?
- What statistical tools are needed to interpret results?

SLIDE 8.57 Triple C Systems Integration—Step 1: Plan.

Step 2: Do (Triple C is Applicable Here)

A plan is just a plan until it is implemented

- Perform activities in accordance with the
  established plan
- USE CPM and PERT tools to determine what to do
- Conduct pilot implementation of the plan

SLIDE 8.58 Triple C Systems Integration—Step 2: Do.

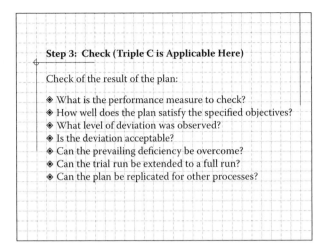

SLIDE 8.59   Triple C Systems Integration—Step 3: Check.

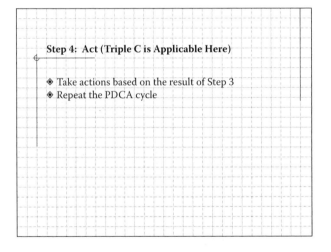

SLIDE 8.60   Triple C Systems Integration—Step 4: Act.

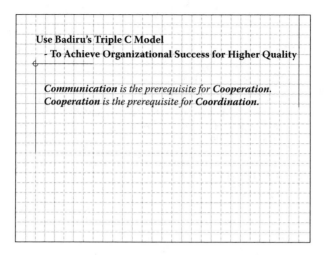

SLIDE 8.61   Triple C Systems Integration—Use Badiru's Triple C Model.

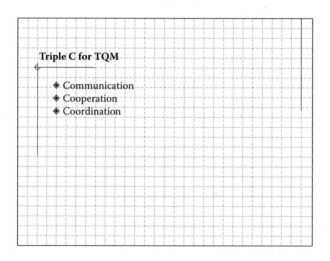

SLIDE 8.62   Triple C Systems Integration—Triple C for TQM.

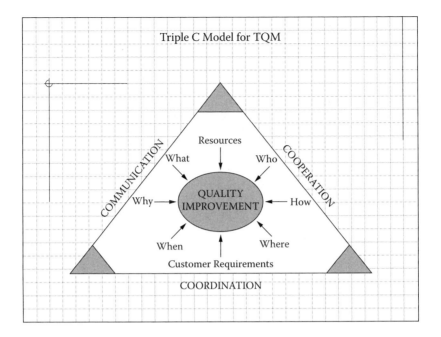

SLIDE 8.63   Triple C Systems Integration—Triple C Model for TQM.

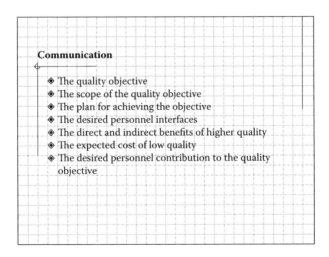

SLIDE 8.64   Triple C Systems Integration—Communication.

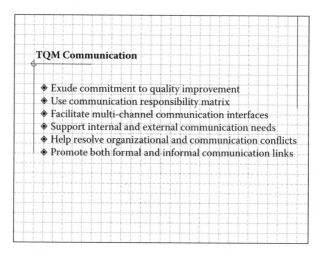

SLIDE 8.65   Triple C Systems Integration—TQM Communication.

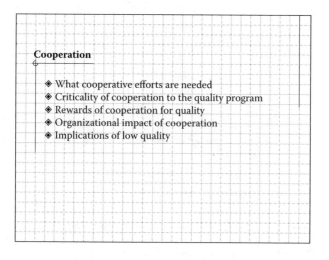

SLIDE 8.66   Triple C Systems Integration—Cooperation.

SLIDE 8.67 Triple C Systems Integration—Coordination.

SLIDE 8.68 Triple C Systems Integration—Example of Responsibility Matrix.

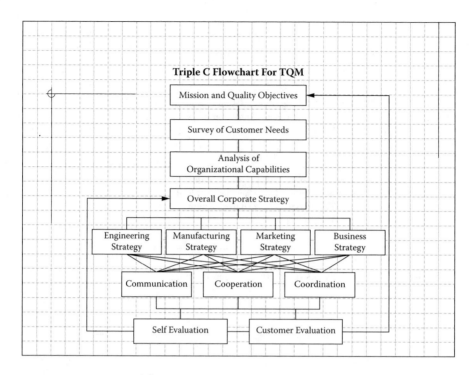

SLIDE 8.69   Triple C Systems Integration—Triple C Flowchart for TQM.

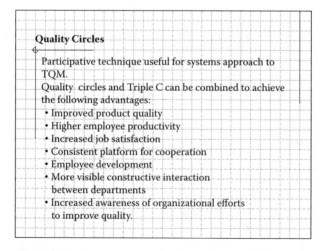

SLIDE 8.70   Triple C Systems Integration—Quality Circles.

ISO 9000 : Quality Standard

Quality objectives can best be achieved by establishing and adhering to standards.

Standard facilitate systems integration.

Isos (Greek for work) → Iso → Equality

ISO 9000 is intended to convey the idea of the <u>invariance</u> that is possible when a standard is available

When we have a standard for a process, the process is expected to produce identical or invariant high-quality units of a product.

SLIDE 8.71   Triple C Systems Integration—ISO 9000: Quality Standard.

ISO 9000 : Quality Management & Quality Assurance Standards

Guidelines for use:

This is the road map that provides guidelines for selecting and using 9001, 9002, 9003, and 9004.

A supplementary publication, ISO 8402, provides quality related definitions.

SLIDE 8.72   Triple C Systems Integration.

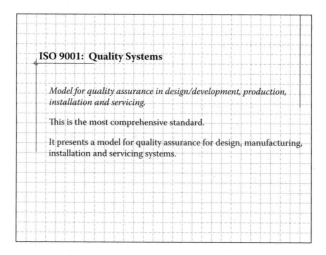

SLIDE 8.73   Triple C Systems Integration—ISO 9001: Quality Systems.

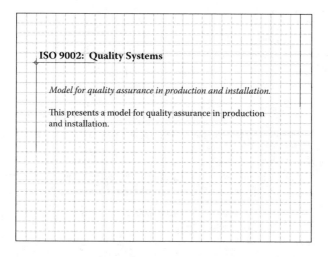

SLIDE 8.74   Triple C Systems Integration—ISO 9002: Quality Systems.

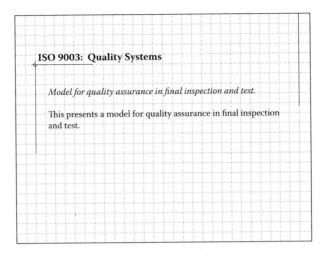

SLIDE 8.75   Triple C Systems Integration—ISO 9003: Quality Systems.

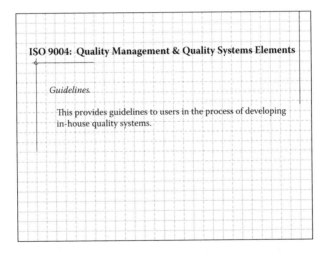

SLIDE 8.76   Triple C Systems Integration—ISO 9004: Quality Management & Quality Systems Elements.

ISO 9000: Audit and Registration

Key Elements of ISO 9000 Audit

◆ A check of whether the company has a documentation process. Does the documentation provide adequate guidelines for workers?

◆ A check of whether everyone in the company is following the documented process. Is everyone aware of updates and changes to the documentation?

◆ A check of how materials are selected. Are appropriate materials selected for specific processes?

◆ A check of how in-house inspection of suppliers' deliveries. Is the company getting what it wants from suppliers?

◆ A check of the calibration and metrology processes. Are calibrations done properly? Are measurements being made accurately?

◆ A check of the procedure for taking corrective actions. Are avenues available for identifying, reporting, and correcting problems?

◆ A check of the internal self-auditing process. Are problems overlooked when they are identified? Is there a formal process review policy? Is the company defensive about obvious quality problems?

SLIDE 8.77   Triple C Systems Integration—ISO 9000: Audit and Registration.

A Vendor Rating Technique

A formal system for vendor rating can be useful in encouraging vendor involvement in TQM efforts.

Requirements:

1. Form a *vendor quality rating team* of individuals who are familiar with the company's operations and the vendor's products.
2. Determine the set of vendors to be included in the rating process.
3. Inform the vendors of the rating process.
4. Each member of the rating team should participate in the rating process.
5. Each member will submit an anonymous evaluation of each vendor based on specified quality criteria.
6. Develop a weighted evaluation of the vendors to arrive at overall relative weights.

SLIDE 8.78   Triple C Systems Integration—A Vendor Rating Technique.

Steps:

1. Let T be the total points available to vendors.
2. $T = 100(n)$, where n = number of individuals in the rating team.
3. Rate the performance of each vendor on the basis of specified quality criteria on a scale of 0 to 100.
4. Let X_{ij} be the rating for Vendor i by team member j.
5. Organize the ratings by team member j as shown below:

    Rating for Vendor 1: X_{1j}
    Rating for Vendor 2: X_{2j}
    .   .
    .   .
    .   .
    Rating for Vendor n: $X_{n,j}$
    Total Rating Points: <u>100n</u>

6. Tabulate the team ratings as shown below and calculate the overall weighted score for each vendor i from the expression below:

$$w_i = \frac{1}{n}\sum_{j=1}^{n} X_{ij}$$

SLIDE 8.79 Triple C Systems Integration—Steps.

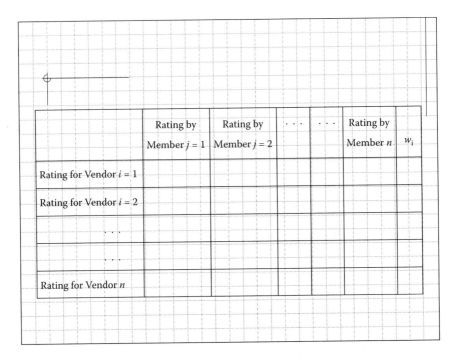

	Rating by Member $j = 1$	Rating by Member $j = 2$	· · ·	· · ·	Rating by Member n	w_i
Rating for Vendor $i = 1$						
Rating for Vendor $i = 2$						
· · ·						
· · ·						
Rating for Vendor n						

SLIDE 8.80 Triple C Systems Integration—Steps 2.

For the case of multiple vendors for the same item, the relative weights, w_i, may be used to determine what fraction, F_i, of the total supply should be obtained from each vendor.

$$F_i = (w_i)(\text{Size of total supply}),$$

Advantages of good vendor rating system:

- Vendor and producer have a joint understanding of customer requirements.
- Skepticism about a vendor's supply is removed.
- Excessive inspection of a vendor's supply is avoided.
- Cost of inspecting a vendor's supply is reduced.
- Vendors reduce their cost by reducing scrap, rework, and returns.
- Vendor morale is improved by the feeling of participation in the producer's mission.

SLIDE 8.81   Triple C Systems Integration—Steps 3.

Concluding Remarks

- Management should address quality problems by instituting a systems approach to total quality management.
- Efforts should be made to continuously improve quality.
- Using TQM approach, the overall business environment should be surveyed for areas where further improvements can be achieved.
- The combination of minor improvements made here and there can lead to a much larger gain in quality improvement.
- Old bad habits and hype should be abandoned in favor of real quality improvements.
- Short-term goals of quality improvement should be extended to long-term and permanent improvement strategies.
- Writing and speaking about quality improvement is not enough.
- Something tangible must be done about it.
- Priorities must be reorganized so that quality is first.
- Quantity and schedule, while very important, should not be allowed to preempt quality.

SLIDE 8.82   Triple C Systems Integration—Concluding Remarks.

CASE PRESENTATION 4: APPLICATION OF TRIPLE C TO CONFLICT RESOLUTION

This section illustrates the application Triple C approach to conflict resolution in the work place. Conflicts develop frequently in the work place due to a variety of factors. Most of the factors are human related. Conflicts can develop as a result of how one person perceives or interprets the actions of another person. Many conflict situations can be resolved through direct communication, cooperation, and coordination of efforts. This is exactly what Triple C does. In this particular case example, organizational conflicts develop in connection with annual appraisal exercise and performance reports.

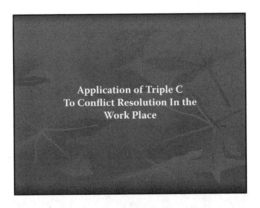

SLIDE 8.83   Conflict Resolution Using Triple C Approach.

Conflict Resolution In The Work Place

> Symptoms: Complaints on appraisal results
> Applying the Triple "C" rule, we shall:
>> communicate the problem and implications explicitly to affected staff and management, elicit patience by staff.
>> Seek cooperation by management coordinate the review to ensure equity and fair play.
>> Listing of interactions that would transform the symptom into the required motivated workforce.

SLIDE 8.84   Conflict Resolution Using Triple C Approach—Conflict Resolution in the Work Place.

SLIDE 8.85   Conflict Resolution Using Triple C Approach—Conflict Resolution in the Work Place 2.

SLIDE 8.86   Conflict Resolution Using Triple C Approach—Conflict Resolution in the Work Place 3.

SLIDE 8.87   Conflict Resolution Using Triple C Approach—Conflict Resolution in the Work Place 4.

Responsibility Matrix

Responsibilities	Affected Staff	Personnel Depart	Management	Review Committee
			ACTORS	
Communicates protest	R			
Acknowledges receipt		R		
Collates all protest	C	R		
Communicates to management		R		
Management setsup Review Committee			R	
Committee reviews protest	C	C		R
Committee interviews affected staff				R
Report submission		C		R
Management takes decision			R	
Personnel department collates results		R		
Personnel department communicates to staff	C	R		

KEY

R = Responsibility

C = Contributors

SLIDE 8.88 Conflict Resolution Using Triple C Approach—Responsibility Matrix.

Resource Allocation For Enhanced Productivity

➤ Symptoms: Inadequate working tools and vehicles.
➤ Applying the Triple "C" rule, we shall communicate the problem explicitly to relevant officers, elicit their cooperation and coordinate the acquisition process by effective follow-up.

SLIDE 8.89 Conflict Resolution Using Triple C Approach—Resource Allocation for Enhanced Productivity.

SLIDE 8.90 Conflict Resolution Using Triple C Approach—Resource Allocation for Enhanced Productivity 2.

SLIDE 8.91 Conflict Resolution Using Triple C Approach—Resource Allocation for Enhanced Productivity 3.

Responsibility Matrix

Responsibilities	User	User HOD	Actors: Supplier HOD	Schedule Officer	Vendor	Budget Officer
Listing of needed resources	R					
Approval and forwarding	C	R				
Authorization and processing			R	R		
Follows-up	R	C				
Obtains vendor's proforma invoice			R	R	R	
Secures budget				R		C
Follows-up			C			
Places and receives order	R			R	C	
Delivers equipment to user	C	C		R		

<u>KEY</u>

R = Responsibility

C = Contributors

SLIDE 8.92 Conflict Resolution Using Triple C Approach—Responsibility Matrix.

Index